T0074194

Physical Chemistry in a Nutshell

Jakob SciFox Lauth

Physical Chemistry in a Nutshell

Basics for Engineers and Scientists

 Springer

Jakob SciFox Lauth ⓘ
FH Aachen
Institut für Angewandte Polymerchemie
Jülich, Germany

ISBN 978-3-662-67636-3   ISBN 978-3-662-67637-0   (eBook)
https://doi.org/10.1007/978-3-662-67637-0

This Springer imprint is published by the registered company Springer-Verlag GmbH, DE, part of
Springer Nature.
The registered company address is: Heidelberger Platz 3, 14197 Berlin, Germany

Preface

The supreme goal of all theory is to make the irreducible basic elements as simple and as few as possible without having to surrender the adequate representation of a simple datum of experience.—(Albert Einstein (1879–1955))

Over the last 150 years, there have been numerous innovations in media technology. Each of these innovations was said to have a significant impact on teaching. The cheers were usually in the majority ("E-learning is a revolution in teaching"), but there were also critical voices ("Are videos spoiling teaching?").

Advances in technology do not automatically make teaching better. It is an illusion to believe that with the right form of teaching or the right medium, learning becomes "easy". Learning is and will remain work in the 21st century. But the instructor can help students do this by generating enthusiasm, showing ways to engage effectively with the material, and setting interesting challenging tasks.

I have been teaching *Physical Chemistry* in higher education for over 25 years and have experimented with both classical and digital technologies. Over time, this has resulted in a multimedia course that is both consistent with student learning (as I gather from evaluations) and compatible with my character and temperament as a lecturer.[1]

In this book, I describe the way I prepare physical chemistry for the younger generation so that they can use it and build on it. The lecture material was divided into many small bites (*lecture bites*); a video of no more than 10 minutes was produced for each of these bites—a format that digital natives, in my experience, use frequently.

The preparation for each of the 12 classroom sessions (I like to call them workshops) is digital and asynchronous (*inverted classroom*). There is an extended online quiz (multiple choice, arithmetic questions, arrangement questions, etc.) to

---

[1] My approach to teaching can be described as *humble teaching*. On the one hand, it is characterized by respect for the great achievements of numerous scientists. The textbook knowledge of today represents the quintessence of centuries of these achievements. My own contribution to this knowledge is practically zero. On the other hand, I am aware that the textbook knowledge of today (as well as much of our current worldview) will be laughed at best in 100 years. Scaled from the future, even many of today's supposed experts will end up very close to *Mount Stupid* on the *Dunning-Kruger* chart.

test prior knowledge. For each workshop, there is a cartoon by Faelis (www. Füchsin.de) as a thumbnail, in which an important fact is humorously presented.

In the workshops themselves, I deliberately do not use digital elements. For example, there is always an *experiment of the day* (mostly performed by students; remains very well memorized) and a *question of the day*. Furthermore, I do not use any presentation programs, but consistently blackboard & chalk (or marker & whiteboard). I work out important facts together with the students on the blackboard.

More details about my teaching can be found in my teaching portfolio.

I wish the reader success and much enjoyment in studying this book. I welcome suggestions for improvement and factual comments.

My thanks go especially to my husband Grey and my friend Faelis for their support and for making the drawings and to the staff of Springer-Verlag for their good cooperation.

Jülich, Germany   Jakob SciFox Lauth
August 2023

Contents

4.8   How Much Enthalpy Is Present in a Molecule
        or in a Chemical Bond?.   58
4.9   How Does Enthalpy Change During a Reaction?.   59
4.10   What Is Entropy? (Factsheet: Table 4.2).   60
4.11   How Can We Measure Entropy?.   60
4.12   When Does Entropy Change?.   62
4.13   How Does Entropy Change in a Reaction?.   62
4.14   How Do We Obtain Free Enthalpy as a Measure
        of Affinity Using the Laws of Thermodynamics?.   63
4.15   When Does the Free Enthalpy Change?.   64
4.16   How Does Free Enthalpy Change During a Reaction?.   65
4.17   How Do We Classify a Process Thermodynamically?.   65
4.18   Summary.   66
4.19   Test Questions.   67
4.20   Exercises.   68

5   Chemical Equilibrium.   69
    5.1   Motivation.   69
    5.2   How Do We Quantify the Location of the Equilibrium?.   69
    5.3   How Do We Classify a Process with Thermodynamic
            Parameters?.   70
    5.4   Is Energy with Us?.   71
    5.5   Is Entropy with Us?.   72
    5.6   How Do We Calculate the Standard Impetus
            (Standard Affinity)?.   72
    5.7   Is Free Enthalpy with Us?.   73
    5.8   How Do We Formulate the Thermodynamic Equilibrium
            Constant?.   73
    5.9   How Do We Calculate the Thermodynamic Equilibrium
            Constant?.   75
    5.10   How Do We Classify a Process in an Entropy/Enthalpy
            Diagram?.   76
    5.11   How Does Temperature Change Standard Impetus and
            Equilibrium Constant?.   77
    5.12   How Can We Change the Position of an Equilibrium?.   78
    5.13   How Can We Provoke Endergonic Reactions?.   79
    5.14   Summary.   80
    5.15   Test Questions.   81
    5.16   Exercises.   82

6   Vapor Pressure.   83
    6.1   Motivation.   83
    6.2   What Is Vapor Pressure?.   83
    6.3   When Are Two Phases in Equilibrium?.   85
    6.4   What Factors Do Affect Vapor Pressure?.   86

Changes of State

1

1.1 Motivation

In this chapter, we learn about the perspective of physical chemistry and, in particular, thermodynamics on the world. What do thermodynamicists mean, when they talk about systems, state variables, process variables, and the like? A firm understanding of these basic terms is important for working through the remaining chapters. (The *motivational picture* of this chapter, Fig. 1.1, illustrates a thermodynamic view on the possible states of pure water).

1.2 How Does Thermodynamics Describe the Oxyhydrogen Reaction?

This chapter is about systems and processes, that is, how a thermodynamicist sees the world.

Using the well-known oxyhydrogen reaction as an example, we will discuss this view.

The reaction of hydrogen with oxygen to form water is formulated by a chemist in this way:

$$2\,H_2\,(g) + O_2\,(g) \rightarrow 2\,H_2O\,(l) \tag{1.1}$$

This is by far not sufficient for physical chemistry to describe the process completely. The aim here is to describe processes unambiguously with numbers, so that every expert can reproduce the experiment exactly and unambiguously.

For example, the complete description could look like Fig. 1.2 and Table 1.1:

We will gradually learn about the physical quantities in this set.

© The Author(s), under exclusive license to Springer-Verlag GmbH, DE, part of
Springer Nature 2023
J. S. Lauth, *Physical Chemistry in a Nutshell*,
https://doi.org/10.1007/978-3-662-67637-0_1

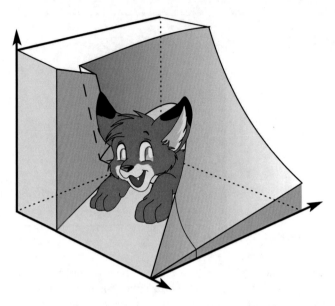

Fig. 1.1 How does thermodynamics see the world? (https://doi.org/10.5446/45974)

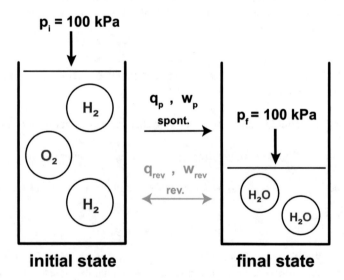

Fig. 1.2 Description of processes using process variables

Table 1.1 Detailed thermodynamic description of initial and final state of a process (oxyhydrogen reaction)

$V_i = 0.0744 \ m^3$	$V_f = 0.000036 \ m^3$
$T_i = 298.15 \ K$	$T_f = 298.15 \ K$
$p_i = 100 \ kPa$	$p_f = 100 \ kPa$
$H°_i = 0 \ kJ$	$H°_f = -572 \ kJ$
$S°_i = 466 \frac{J}{K}$	$S°_f = 140 \frac{J}{K}$
$G°_i = 0 \ kJ$	$G°_f = -474 \ kJ$

Fig. 1.3 Container with oxyhydrogen gas as thermodynamic system

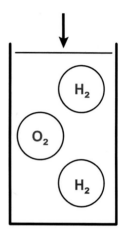

1.3   How Do We Describe the State of a System?

Let us start with the reactants (also called reactants or substrates):

2 moles of hydrogen and 1 mole of oxygen are in a container, which is called a "system" by physical chemistry (illustrated in Fig. 1.3).

A system is a defined part of the universe, delimited by boundaries from its surroundings.

Our system here must now be described unambiguously and completely. This description is done with the so-called state variables, which quantify the system.

Many of these state variables are known from everyday life, such as mass, temperature, or pressure.

Thermodynamics "invents" (or rather "defines") a number of other state variables that quantify primarily the energetic and energy-like aspects of the system. The most important of these quantities are the enthalpy H, the entropy S, and the free enthalpy ("*GIBBS ENERGY*") G.

H, S, and G are properties of the system—just like the mass or the volume. But with these quantities we can formulate and apply general laws, which are valid in thermodynamics ("Laws of Thermodynamics") in a simple form. We find more about this in Chap. 4: "Are energy and/or entropy with us?".

1.4   How Many Variables Do We Need to Completely Specify a System?

How many numbers do we need to describe our system unambiguously? *Josiah Gibbs*—one of the great scientists in thermodynamics—established a rule for this.

Actually, it is even a law—the simplest law of thermodynamics:

$$F = C - P + 2 \tag{1.2}$$

This is *GIBBS'* phase rule. *GIBBS* states that for a number C of components and a number P of phases exactly F intensive quantities are needed to describe the system unambiguously. In our example, we have two components (hydrogen and oxygen), i.e. two different types of particles: $C = 2$.

Furthermore, our example system is homogeneous, i.e. there is only one phase (phase = homogeneous area in the system), namely a gas phase. The simple calculation gives $F = 2-1+2 = 3$. We have got three degrees of freedom; we need exactly three parameters to describe the system unambiguously—not less and not more. We can choose these three state variables, e.g. temperature T, molar volume $\overline{V}$, and indication of the composition x. We are "free" in the choice of the quantities, we could also choose the density, the pressure, and the enthalpy; however, there must be exactly three parameters.

The product water—again a system—consists of only one component, H_2O. It is also homogeneous, so it consists of one phase (liquid). We need two ($F = 1-1+2 = 2$) parameters for complete description. The specification of temperature T and molar volume $\overline{V}$ is sufficient to quantify the system "water" unambiguously (just like any other pure substance).

All other state variables—all other properties—then adjust "automatically." If T and $\overline{V}$ are fixed, all other quantities (pressure, enthalpy, etc.) are also fixed.

Therefore, all states of a pure substance can be represented graphically on a plane (e.g., the T-$\overline{V}$ plane) as points.

We get even more information from a three-dimensional phase diagram, in which, for example, the pressure p is listed as additional information.

The $p\overline{V}\,T$ phase diagram of a pure substance ("one-component system") will be our guide through much of this lecture series (see Fig. 1.4).

1.5   How Do We Describe a Process with Thermodynamic Quantities?

Back to our oxyhydrogen reaction. We have now completely described our reactants (substrates) and our product thermodynamically and now we want to discuss the actual reaction—the process from initial state i ("initial") to final state f ("final"). A process is always a change of state—in this case, a chemical change of state. A process can be described by the change of the state variables ΔZ.

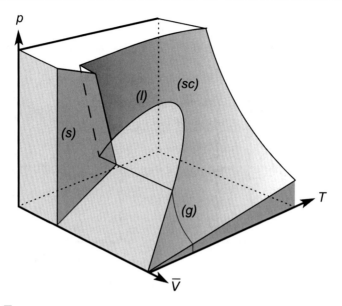

Fig. 1.4 $p\overline{V}T$ phase diagram of water [(H$_2$O), one-component system]. s solid, l liquid, g gaseous, sc supercritical

We calculate the difference between the final value Z_f and the initial value Z_i of the state variable.

$$\Delta Z = Z_f - Z_i \tag{1.3}$$

For example, we can define the volume change as

$$\Delta V = V_f - V_i \tag{1.4}$$

In our example, the volume change is negative; it is a so-called exochoric process.

Of course, a state variable can also remain constant, e.g. in our example the pressure remains constant (see Fig. 1.5).

This is an isobaric process.

$$p_f = p_i \tag{1.5}$$

$$\Delta p = p_f - p_i = 0 \tag{1.6}$$

Let us take a closer look at the "new" state variables of thermodynamics.

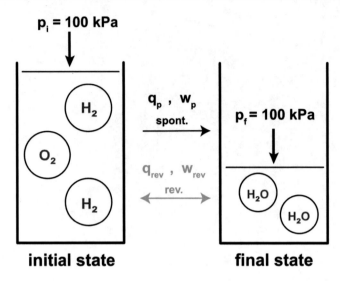

Fig. 1.5 Isobaric isothermal formation of water from oxyhydrogen gas as a thermodynamic process

---

1.6   How Does Energy Change During a Process?

The enthalpy H is a measure of how much energy is in a system. In the appendix, we find a detailed thermodynamic table, in which this quantity is tabulated for pure substances in the standard state ($p° = 100$ kPa, mostly 25 °C) (the so-called standard enthalpies of formation.$\Delta_f H°$). For the substances involved in our reaction, we find the following values:

$$H_2O(l): \Delta_f H° = -285.8 \frac{kJ}{mol} \tag{1.7}$$

$$H_2(g): \Delta_f H° = 0 \frac{kJ}{mol} \tag{1.8}$$

$$O_2(g): \Delta_f H° = 0 \frac{kJ}{mol} \tag{1.9}$$

Hydrogen and oxygen both have the enthalpy 0, i.e. our reactant oxyhydrogen also has the enthalpy 0 in total. The product water, on the other hand, is significantly lower in energy: −572 kJ for 2 mol of water.

So in the process the energy decreases, it is an exothermic process. The standard enthalpy of reaction $\Delta_r H°$ is negative.

$$\Delta_r H^\circ = H_f - H_i \tag{1.10}$$

$$\Delta_r H^\circ = -572 \text{ kJ} \tag{1.11}$$

The standard enthalpy of reaction is a measure of how the energy changes in a process. In our case, the product (2 moles of water) is 572 kJ lower in energy than the reactants.

1.7 How Does Chaos (Entropy) Change During a Process?

Entropy S is another thermodynamic quantity and it is somewhat more obscure than enthalpy. S is a quantitative measure of disorder, of chaos, of the lack of information ("neginformation"). In the appendix, we find a detailed thermodynamic table, in which also this quantity is tabulated for pure substances in the standard state ($p^\circ = 100$ kPa, mostly 25 °C) (the so-called standard entropies S°). For the substances involved in our reaction, we find the following values:

$$H_2O(l) : S^\circ = 69.9 \frac{J}{\text{mol K}} \tag{1.12}$$

$$H_2(g) : S^\circ = 130.6 \frac{J}{\text{mol K}} \tag{1.13}$$

$$O_2(g) : S^\circ = 205 \frac{J}{\text{mol K}} \tag{1.14}$$

The pure sum of the entropy of the reactants is 466 J/K. This is quite a high entropy, but not unusual for "chaotic" gases (in fact, the words "gas" and "chaos" have the same root word).

In fact, the entropy of the oxyhydrogen gas is even higher, since the chaos increases further when two pure substances are mixed.

Liquid water, on the other hand, is much more ordered with an entropy of 140 J/K. Thus, in our process, the entropy decreases; the standard reaction entropy $\Delta_r S^\circ$ is negative; it is a so-called exotropic process.

$$\Delta_r S^\circ = S_f - S_i \tag{1.15}$$

$$\Delta_r S = -326 \frac{J}{K} \tag{1.16}$$

The specification of the standard enthalpy of reaction $\Delta_{rxn} H^\circ$ and the standard reaction entropy $\Delta_{rxn} S^\circ$ is typical for the thermodynamic approach to a process.

The basic laws of thermodynamics (first law, second law) can be applied in a simple way if we know $\Delta_r H^\circ$ and $\Delta_r S^\circ$.

1.8   How Does Instability Change During a Process?

The thermodynamic quantity G is particularly important. G is the free enthalpy (or "GIBBS' energy"), a measure of instability. In the appendix, we find a detailed thermodynamic table in which this quantity is tabulated for pure substances in the standard state ($p° = 100$ kPa, usually $25\ °C$) (the so-called standard free enthalpy of formation $\Delta_f G°$ or also called "standard chemical potential" $\mu°$). For the substances involved in our reaction, we find the following values:

$$H_2O(l): \Delta_f G° = -237.13\frac{kJ}{mol} \tag{1.17}$$

$$H_2(g): \Delta_f G° = 0\frac{kJ}{mol} \tag{1.18}$$

$$O_2(g): \Delta_f G° = 0\frac{kJ}{mol} \tag{1.19}$$

The sum of the instability of the reactants is 0; the instability of 2 mol water is -474 kJ. This means: water is much more stable than oxyhydrogen. During the reaction the instability decreases; the free standard enthalpy of reaction $\Delta_{rxn} G°$ is negative; this is a so-called exergonic process.

$$\Delta_r G° = \Delta_f G°_{final} - \Delta_f G°_{initial} \tag{1.20}$$

$$\Delta_r G° = -474\ kJ \tag{1.21}$$

The standard free enthalpy of reaction $\Delta_r G°$ is also called "standard impetus" or "standard affinity" of a process.

Endergonic processes (positive $\Delta_r G°$) can never run spontaneously from reactants to products (complete conversion).

We can therefore describe a process thermodynamically by describing both initial state and final state with state variables. Usually, the important thermodynamic state variables are also included: enthalpy, entropy, and free enthalpy.

We know the energy quantities heat q and work w from everyday life, but these are to be understood thermodynamically a little differently.

1.9   What Is the Sign of the Process Variables Heat and Work?

Heat and work are forms of energy exchange between system and surroundings. The sign convention of thermodynamics applies here.

Quantities of energy absorbed by the system are evaluated positively; quantities of energy released by the system are evaluated negatively.

Figure 1.6 shows an example where the system absorbs work; then the work has a positive magnitude: $w > 0$.

Fig. 1.6 Sign convention in thermodynamics

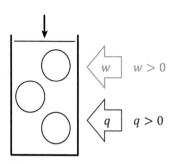

In the same figure, the system also absorbs heat; the heat also has a positive magnitude (the so-called endothermic process).

In many chemical reactions, heat is released from the system; these are then exothermic processes.

For example, when we heat a cup of water (125 g) from 25 °C to 75 °C, our system (water) absorbs energy in the form of heat from the surroundings; it is an endothermic process.

In this example, heat changes temperature, we talk about "sensible heat."

1.10 How Do We Measure Heat?

We could measure the sensible heats by using the well-known equation from physics:

$$q_{sen} = C\Delta T \tag{1.22}$$

C is the heat capacity, ΔT is the temperature difference (final temperature − initial temperature).

If heat capacity is not constant, we have to modify the equation.

$$q_{sen} = \int C\,dT \tag{1.23}$$

The specific heat capacity of liquid water is 4.184 J/(K g) (1 cal/(K g)). 125 g of water thus has a heat capacity of 523 J/K. Using these values, we can calculate the sensible heat.

$$q_{sen} = C\Delta T = 523\frac{J}{K}\,(50\text{ K}) = 26\text{ kJ}$$

We need 26 kJ of heat to heat 125 g of water from 25 °C to 75 °C.

When adding heat to the system results in no change in temperature, the heat is referred to as "latent heat."

Figure 1.7 summarizes the processes that occur when 1.0 g of water is heated from the solid state (273 K) to the gaseous state (373 K and beyond).

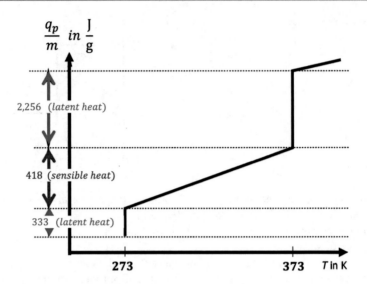

$\dfrac{q_p}{m}$ *in* $\dfrac{J}{g}$

2,256 (*latent heat*)

418 (*sensible heat*)

333 (*latent heat*)

273   373   *T* in K

Fig. 1.7 Sensible and latent heat with heating up water

If we add heat to 1.0 g of ice at 273 K, then initially the temperature remains constant, until the ice has melted. We need 333 J to completely melt 1 g of ice. This is latent heat—the heat of fusion $\Delta_{fus}q$ (or melting enthalpy $\Delta_{fus}H$).

With further addition of heat, the temperature of the now liquid water increases up to 100 °C. Here we can use the formula for sensible heat. Then we add latent heat again to evaporate the water (heat of evaporation $\Delta_{vap}q$ or enthalpy of evaporation $\Delta_{vap}H$).

The heating of the water vapor then corresponds again to a sensible heat.

Liquid water has a relatively large heat capacity (large slope in Fig. 1.7); the heat capacity of ice or water vapor is only about half as large.

1.11   How Do We Measure Work?

From physics, we know the equation for calculating work:
Work is force times distance.

$$w = F \cdot s \tag{1.24}$$

In general, work is always the scalar product of a "force-like quantity" and a "path-like quantity."

Electrical work can be calculated using voltage U, current I, and time t, for example.

$$w_{el} = U \cdot I \cdot t \tag{1.25}$$

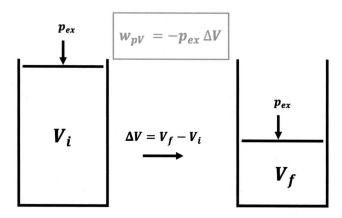

Fig. 1.8 Work done when volume is reduced by an external pressure

If we charge our smartphone on a standard USB port for 3 h, the voltage is 5 V and 1 A of current flows.

Accordingly

$$w_{el} = 5\ \text{V} \quad 1\ \text{A} \quad 10\ 800\ \text{s} = 54\ \text{kJ} \tag{1.26}$$

54 kJ of work is transferred to the cell phone's battery. Converted, this corresponds to the heat requirement of about two cups of tea.

In thermodynamics, pressure–volume work is particularly important. Whenever the volume of a system changes—becomes smaller or larger—work is involved.

Figure 1.8 illustrates this. The atmospheric pressure p_{ex} creates a force on our system. If we now change the volume of the system (e.g., in the oxyhydrogen reaction), we have a "path" along which the force acts. We get for this so-called pressure–volume work w_{pV}

$$w_{pV} = -p\Delta V \tag{1.27}$$

Our 74.4 L of oxyhydrogen reacts isobarically to 36 mL of water. The reaction is exochoric with $\Delta V = -74.4$ L. At an external pressure of 100 kPa this means a volume work of

$$w_{pV} = -(100\ \text{kPa})(-74.4\ \text{L}) = 7.4\ kJ \tag{1.28}$$

The atmosphere does 7.4 kJ of work on the oxyhydrogen—for free, so to speak.

1.12 How Do We Describe a Process Thermodynamically?

On the one hand, a process is defined by its initial and final state (quantified by the respective state variables Z_i and Z_f); on the other hand, the specification of the process variables work w and heat q is also relevant for a process.

It is particularly important that the process variables depend on the path.

The changes of the state variables from Z_i to Z_f are always the same, no matter what path we take. Whether we ignite the oxyhydrogen directly or let it react slowly in a fuel cell, the changes of the state variables ΔZ are the same.

The numbers for work and heat, on the other hand, can be different for the oxyhydrogen reaction depending on path. Pressure–volume work is always 7.4 kJ, because we always have the same volume change. But if we carry out the reaction directly spontaneously (in an oxyhydrogen burner) we get no more additional work—the useful work $w_{use} = 0$, but we get a lot of heat: -572 kJ.

If we carry out the same process in a fuel cell, then our system supplies a lot of electrical work, namely -474 kJ—but correspondingly less heat, only -98 kJ.

Because of the path dependence, we should always mark the process variables w and q with an index which specifies the path.

1.13 Summary

A thermodynamicist investigates systems-defined parts in the universe. Thermodynamics describes the state of these systems with state variables and also invents new state variables (enthalpy, entropy, and free enthalpy).

The change of states is called "process" by a thermodynamicist. A process is described by "Δ-sizes"—ΔT, Δp, etc., and by the process variables heat q and work w, which are path-dependent.

We must therefore always specify the path.

Important equations are the GIBBS phase rule

$$F = C - P + 2 \tag{1.29}$$

The equation for calculating the sensible heat.

$$q_{sen} = C\Delta T \tag{1.30}$$

and the equations for calculating the electric work and the pressure–volume work.

$$w_{el} = U \cdot I \cdot t \tag{1.31}$$

$$w_{pV} = -p\Delta V \tag{1.32}$$

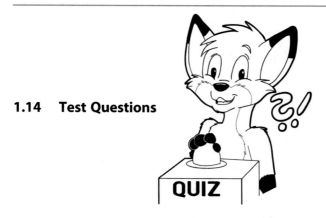

1.14   Test Questions

1. In a container, 1 kg of ice, 1 kg of water and 10 g of water vapor are in equilibrium. How many degrees of freedom does the system have?
2. The gas in an air pump (=system) is reversibly compressed isothermally. What are the signs for the process variables w and q?

3. 1 liter of water (=system) is heated from 20 °C to 100 °C in a kettle. What are the signs of the process variables heat q and work w?
4. The oxyhydrogen reaction ($2H_2+O_2 \rightarrow 2H_2O$) is carried out twice at constant temperature and pressure.
   In one case, the reaction is spontaneous (oxyhydrogen burner) and heat q_p is released; in the other case, it is reversible (fuel cell) and heat q_{rev} is released.
   Mark the correct answer(s).
   (a) $q_{rev} = q_p$
   (b) $|q_{rev}| < |q_p|$
   (c) $|q_{rev}| > |q_p|$
   (d) $q_{rev} < 0$
5. 1 mol of carbon dioxide gas (=system) is cooled isobarically—starting from standard conditions—until complete solidification. What are the signs of the process variables heat q and work w ?

1.15   Exercises

1. How much heat is required in total to convert at 101.3 kPa
   1 mol of solid water (ice) of −25 °C into
      1 mol of gaseous water (water vapor) at 125 °C?
      Specific heat capacities:
      $c_p(s) = 2.03$ kJ/(kg K); $c_p(l) = 4.18$ kJ/(kg K); $c_p(g) = 1.84$ kJ/(kg K)
      Molar heats of phase transformation: $\Delta_{fus}H = 6.01$ kJ/mol; $\Delta_{vap}H = 40.67$ kJ/mol

2. A polystyrene block (2.00 kg, temperature 50.0 °C, $\langle c_p \rangle = 1.30 \frac{\text{kJ}}{\text{kg °C}}$) is placed
   under isobaric conditions in a water bath (10.0 kg, temperature 20.0 °C,
   $\langle c_p \rangle = 4.18 \frac{\text{kJ}}{\text{kg °C}}$). After some time, thermal equilibrium has been reached.
   Calculate the heat q exchanged.

3. Figure 1.9 shows the phase diagram of carbon dioxide. Mark the following
   process in it:
   Carbon dioxide gas is completely liquefied isothermally by compression—
   starting from standard conditions.

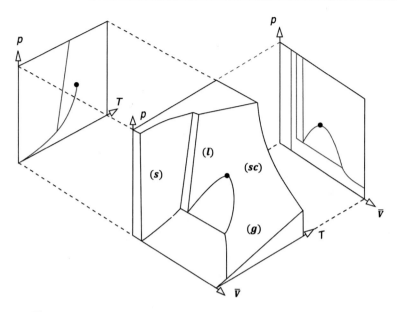

Fig. 1.9 $p\overline{V}T$ phase diagram of carbon dioxide and projection onto the pT and $p\overline{V}$ -plane [($_{CO2}$), one-component system]. s solid, l liquid, g gaseous, sc supercritical

Gases

2

2.1   Motivation

Gases play an important role in nature and in technology.

How can we describe and understand gases both macroscopically and microscopically—that is, from the model presentation? (The *motivational picture* of this chapter, Fig. 2.1, shows all possible thermodynamic states of an ideal gas).

2.2   Where Do We Find "Ideal" and "Real" Gases in the Phase Diagram?

We start the topic "gases" with the phase diagram of a pure substance—our common thread through much of thermodynamics (see Fig. 2.2).

Where do we find the gaseous state here, especially the state of the so-called ideal gas? Ideal gases are those gases whose state is at a due distance from the critical point—we find them in the range of high temperatures and high molar volumes on the phase surface.

In this realm, the phase surface is continuously curved and it can be described very easily, namely by the ideal gas equation. When we move on to real gases—i.e. gases near their critical point—we have to consider deviations from this ideal behavior.

Unlike ideal gases, real gases can be liquefied by applying an appropriately high pressure.

We zoom to the realm of the ideal gas in the phase diagram:

This phase surface can be described mathematically very well by the ideal gas equation

© The Author(s), under exclusive license to Springer-Verlag GmbH, DE, part of
Springer Nature 2023
J. S. Lauth, *Physical Chemistry in a Nutshell*,
https://doi.org/10.1007/978-3-662-67637-0_2

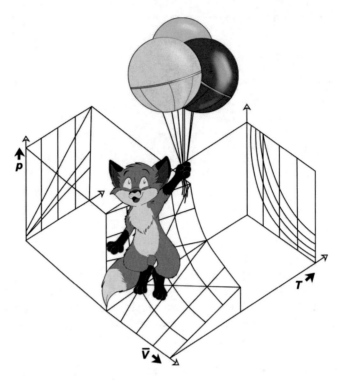

Fig. 2.1 How can we describe gaseous systems macroscopically and microscopically? (https://doi.org/10.5446/45976)

$$p = \frac{RT}{\overline{V}} \tag{2.1}$$

If the ideal gas equation is solved for pressure here; mathematically, it represents a function of two variables. Since the pressure is a state variable, there is a total differential of this function. More details can be found in textbooks of mathematics.

2.3   How Do Gases React on Change of Volume?

The ideal gas law has a long history.

In the seventeenth century, several scientists studied the compressibility of gases; that is, the behavior of a gas when its volume changes (see Fig. 2.4).

It turned out that the behavior of all gases in this respect is very simple: pressure and volume are inversely proportional—a halving of the volume means a doubling of the pressure (provided that the process is isothermal, i.e. the temperature is identical in the initial and final states).

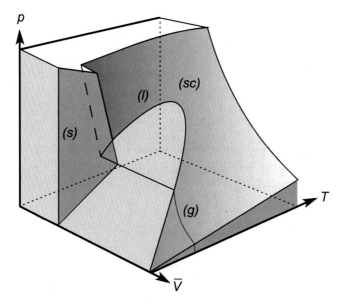

Fig. 2.2 $p\overline{V}T$ phase diagram of water [(H₂O), one-component system]. *s* solid, *l* liquid, *g* gas, *sc* supercritical

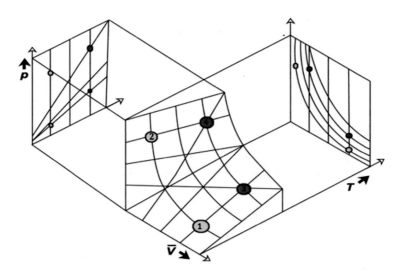

Fig. 2.3 $p\overline{V}T$ phase diagram of an ideal gas

$$p_i \cdot V_i = p_f \cdot V_f \qquad (2.2)$$

This is BOYLE–MARIOTTE'S law. In the phase diagram in Fig. 2.3, the states (1) and (2) marked with bright dots have identical temperature. The isothermal connecting these states obeys BOYLE–MARIOTTE'S law and corresponds to a hyperbola.

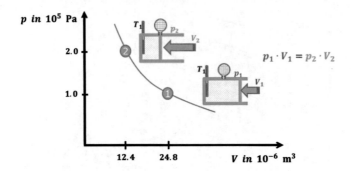

Fig. 2.4 BOYLE–MARIOTTE isotherm

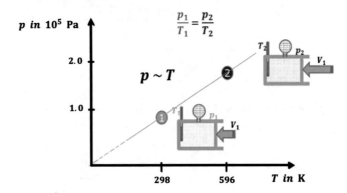

Fig. 2.5 GAY-LUSSAC isochor

States (3) and (4) marked with dark dots lie on another isotherm (hyperbola) at a higher temperature.

2.4   How Do Gases React on Change of Temperature?

What happens when we increase or decrease the temperature of a gas? (see Fig. 2.5)

Again, the behavior of all ideal gases can be described very simply: Volume and absolute temperature are proportional—if we increase the temperature by 10%, the volume also increases by 10% (assuming the process is isobaric, i.e. the pressure is identical in the initial and final states). CHARLE'S law applies.

$$\frac{V_i}{T_i} = \frac{V_f}{T_f} \tag{2.3}$$

Analogously, pressure and temperature are also proportional if the process is isochoric (constant volume). GAY-LUSSAC's law applies.

$$\frac{p_i}{T_i} = \frac{p_f}{T_f} \qquad (2.4)$$

Isobars and isochores are straight lines in the phase diagram. The states (1) and (3) in Fig. 2.3 lie on an isochore, the states (2) and (4) in Fig. 2.3 lie on an isobar.

2.5   How Do Gases React on Change of Amount of Substance?

The influence of the amount of substance on the volume is very simple for gases as for any other substance: the larger the amount of substance, the larger the volume—there is a direct proportionality. In the case of gases, however, there is the additional fact that the molar volume (or mole volume) $\overline{V}$ is identical for all gases.

$$\frac{V}{n} = \overline{V} = \text{const.} \qquad (2.5)$$

This was formulated by AVOGADRO in his hypothesis as follows: Equal amounts of gas occupy equal volumes (precondition: equal temperature and pressure).

We should remember that the volume of any gas is 22.7 L at standard conditions (STP) and 22.4 L at "normal" conditions (DIN1343: 1 atm; 0 °C).

$$\overline{V} = (22.7\,\text{L})_{100\text{kPa};\,0\,°\text{C}} \qquad (2.6)$$

Standard conditions have been defined by IUPAC as 0 °C and 100 kPa and are often abbreviated with the index °.

2.6   How Do We Describe the State of an Ideal Gas?

If we summarize all the laws for ideal gases, we arrive at the ideal gas law, which provides a very good mathematical representation of the phase surface shown in Fig. 2.3:

$$p = \frac{RT}{\overline{V}} \qquad (2.7)$$

or converted

$$pV = nRT \qquad (2.8)$$

The ideal gas constant is the same for all gases:

$$R = 8.314\,\text{J/mol K} \qquad (2.9)$$

Instead of the unit Joule, we may also use the product of liter and kilopascal.

$$R = 8.314 \text{ L kPa/mol K} \qquad (2.10)$$

2.7 How Do We Describe a Mixture of Gases?

The ideal gas law applies to all gases and also to gas mixtures. Dry air is a gas mixture of 78 mol% nitrogen, 21 mol% oxygen, and 1 mol% argon; the air components are represented by different symbols in Fig. 2.6.

In a gas mixture, we can measure the total pressure with a common manometer. All components of the gas contribute to the total pressure. In our example, the total pressure is 100 kPa.

The idea of assigning a so-called partial pressure p_i to each component i goes back to DALTON. p_i is the pressure that would prevail if all components except component i were removed.

According to DALTON, partial pressure can be calculated very simply from the mole fraction y (molar fraction) of the gas component.

$$p_i = y_i \, p_{total} \qquad (2.11)$$

Oxygen has a mole fraction of 0.21 in air, so its partial pressure is 21% of the total pressure, i.e. 21 kPa in our example. Consequently, nitrogen in our example has a partial pressure of 78 kPa.

The measurement of partial pressure requires selective pressure probes. For example, the lambda probe is a selective probe for oxygen.

The partial pressures add up to the total pressure. The partial pressures of the components of the air add up to 100 kPa in our example.

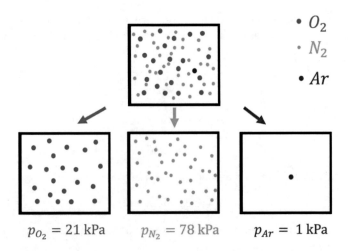

Fig. 2.6 Gas mixture to illustrate the concept of partial pressure according to DALTON

$$p_{total} = \sum_i p_i \qquad (2.12)$$

Air usually also contains between 0 and 2 mol% water vapor (at 20 °C). This means the partial pressure of water in humid air is between 0 and 2 kPa.

2.8   What Energy Do Gas Particles Have?

Let us look at a gas microscopically. The kinetic theory of gases developed by MAXWELL and BOLTZMANN assumes that gases are small particles moving very fast.

The volume of the particles is small compared to the total volume of the gas.

The velocity of the particles changes constantly depending on how often they collide with other gas particles or with the wall (see Fig. 2.7).

Based on these model ideas, MAXWELL and BOLTZMANN were able to calculate that the temperature of a gas is microscopic in the translational motion of the particles. $\langle U_{trans} \rangle$ is the mean kinetic energy of the particles, which is linked to the locomotion (translation). It is the microscopic equivalent of the macroscopic quantity of temperature.

$$\langle U_{trans} \rangle = \frac{3}{2} R T \qquad (2.13)$$

If we calculate this energy for any gas (e.g., argon) at room temperature (20 °C), we get a value of

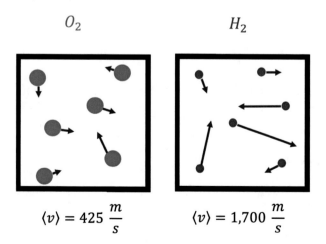

$$O_2 \qquad\qquad H_2$$

$$\langle v \rangle = 425 \ \frac{m}{s} \qquad\qquad \langle v \rangle = 1{,}700 \ \frac{m}{s}$$

Fig. 2.7 Different gases at the same temperature and pressure

$$\langle U_{\text{trans}} \rangle = \frac{3}{2} \; 8.314 \frac{\text{J}}{\text{mol K}} \; 273 \text{ K} = 3.4 \text{ kJ/mol} \tag{2.14}$$

This is the thermal energy contained in 1 mole of argon and, of course, in every mole of any other gas, because there is no substance-specific quantity in the equation—there is only temperature as a parameter. It is useful to know the magnitude of thermal energy at STP, especially to be able to compare it to other important energies. The energy required to evaporate water is 40 kJ/mol, about 10 times higher than the thermal energy at STP; the energy required to split the O-H bond is a factor of 100 higher.

2.9 How Fast Do Gas Particles Move?

The kinetic theory of gases according to MAXWELL and BOLTZMANN provides the following relation for the velocity distribution in a gas:

$$F(v) = 4\pi \left(\frac{M}{2\pi RT} \right)^{1.5} v^2 e^{\left(\frac{Mv^2}{2RT} \right)} \tag{2.15}$$

This distribution function is worth taking a closer look at (see Figs. 2.8 and 2.9).

The velocity of the gas particles is plotted on the x-axis and a measure of abundance is plotted on the y-axis. The curve is not symmetrical; therefore, the average velocity $\langle v \rangle$ is slightly to the right of the maximum. $\langle v \rangle$ can be calculated from the temperature and the molar mass.

$$\langle v \rangle = \sqrt{\frac{8RT}{\pi M}} \tag{2.16}$$

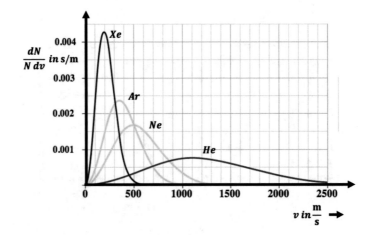

Fig. 2.8 MAXWELL–BOLTZMANN velocity distribution for several noble gases at 298 K

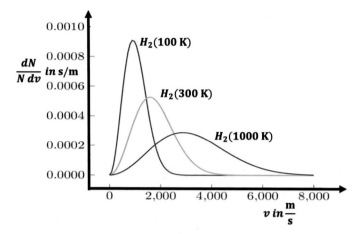

Fig. 2.9 MAXWELL–BOLTZMANN velocity distribution for hydrogen for three temperatures

The higher the temperature and the smaller the molar mass, the faster a particle is on average. For argon $\langle v \rangle$ is 380 m/s under standard conditions.

$$\langle v \rangle = \sqrt{\frac{8RT}{\pi M}} = \sqrt{\frac{8 \cdot 8.314 \frac{J}{mol\,K} \cdot 273\ K}{\pi\ 0.040 \frac{kg}{mol}}} = 380 \frac{m}{s} \tag{2.17}$$

Water as a lighter particle has a significantly higher velocity in the gas phase than argon at the same temperature.

2.10   How Often Do Gas Particles Collide?

Gas particles are very fast by macroscopic standards. However, they do not move very far in a gas because they collide very often with other particles. These collisions are also quantified by the kinetic theory of gases according to MAXWELL and BOLTZMANN.

The gas particles collide with the wall—this is the molecular cause of the pressure. With the following formula, we can calculate the frequency of the wall collisions.

$$\langle z_w \rangle = \frac{1}{4} \langle v \rangle \frac{N_A\, p}{R\,T} \tag{2.18}$$

For the calculation of the frequency of intermolecular collisions, the collision cross section σ is necessary to quantify the bulkiness of a particle

$$\langle z \rangle = \sqrt{2}\,\sigma \langle v \rangle \frac{N_A\, p}{R\,T} \tag{2.19}$$

For argon with an collision cross section of 0.36 nm^2, we calculate at standard conditions an collision frequency of

$$\langle z \rangle = \sqrt{2} \cdot 0.36 \cdot 10^{-18} \text{m}^2 \cdot 380 \frac{\text{m}}{\text{s}} \cdot \frac{6.02 \cdot 10^{23} \frac{1}{\text{mol}} \cdot 100 \text{ kPa}}{8.314 \frac{\text{J}}{\text{mol K}} \cdot 273 \text{ K}} = 5.1 \cdot 10^9 \frac{1}{\text{s}} \quad (2.20)$$

An argon particle in air collides with other gas particles more than 5 billion times in just 1 s!

2.11   What Distance Do Gas Particles Travel Between Two Collisions?

An argon particle in the air is about 400 m/s fast and witnesses about 5 billion collisions per second. Putting these two quantities in relation, we obtain the so-called mean free path $\langle \lambda \rangle$

$$\langle \lambda \rangle = \frac{\langle v \rangle}{\langle z \rangle} = \frac{RT}{N_A \sqrt{2} \sigma p} \quad (2.21)$$

This important parameter indicates how far a particle travels before it collides with another particle. For argon, this parameter is

$$\langle \lambda \rangle = \frac{380 \frac{\text{m}}{\text{s}}}{5.1 \cdot 10^9 \frac{1}{\text{s}}} = 75 \text{ nm} \quad (2.22)$$

75 nm is smaller than the wavelength of light (wavelength of yellow light: 590 nm).

Mean free path depends very much upon pressure, as can be seen from Eq. (2.21). At very low pressure, the mean free path is of the same order of magnitude as the vessel dimensions (1 m). This is important for vacuum technology, e.g. for mass spectrometers.

2.12   How Do We Describe Deviations from Ideal Behavior?

We recall the premises of the kinetic theory of gases: the particles are very small and have no attractive forces on each other.

This applies in the range of the ideal gas, if the temperature is high and the volume is large. But if we let the gas cool down and/or decrease its volume, then the intrinsic volumes of the particles and the attractive forces of the particles come into play and then the gas will no longer behave ideally.

VAN-DER-WAALS quantified these deviations from the ideal behavior; he modified the ideal gas equation and added two correction factors.

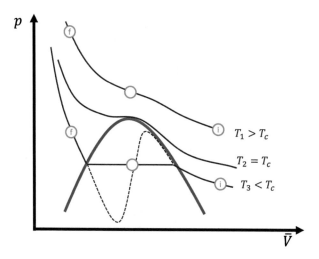

Fig. 2.10 Isotherms of CO_2 above and below the critical point (with the so-called VAN-DER-WAALS-loop below T_c)

$$\left(p + \frac{a}{\overline{V}^2}\right)\left(\overline{V} - b\right) = RT \tag{2.23}$$

The factor a is a measure of the attractive forces of the particles. The so-called internal pressure $\frac{a}{\overline{V}^2}$ is added to the real pressure, because the measured pressure p is smaller than the ideal pressure.

The factor b is called "covolume" and is a measure of the intrinsic volume of the particles; b is subtracted from the measured volume because the ideal volume available to the particles is smaller than the real volume. In the ideal gas, of course, both a and b are equal to 0.

If we compare the real gases argon and water (see Appendix), water has an internal pressure factor a four times larger than argon ($554 \frac{kPa\,L^2}{moL^2}$ compared to $138 \frac{kPa\,L^2}{moL^2}$); thus, significantly larger intermolecular attractive forces act in water.

With respect to covolume b, however, there are hardly any differences between the two particles ($0.031 \frac{L}{mol}$ for water and $0.032 \frac{L}{mol}$ for argon).

We can obtain the VAN-DER-WAALS-factors a and b from the critical quantities. In fact, a "VAN-DER-WAALS-gas" shows only 3/8 of the behavior of an ideal gas at its critical point (see Fig. 2.10).

$$p_c\,\overline{V}_c = \frac{3}{8}R\,T_c \tag{2.24}$$

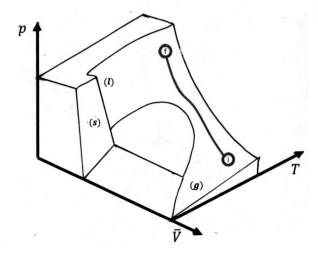

Fig. 2.11 Isothermal compression of water vapor above the critical point

2.13   What Happens When a Gas Is Compressed Above Its Critical Temperature?

We compress a gas above its critical temperature.

Above the critical temperature, there are no two-phase regions. Figure 2.11 shows the compression of water vapor (critical temperature: 374 °C) at 400 °C as an example. However, we could also compress a so-called permanent gas such as methane (critical temperature: −82.6 °C) at room temperature.

In any case, during supercritical compression, the density of the gas will keep increasing just like its pressure, but nothing else happens.In particular, the formation of another phase never occurs. Liquefaction above the critical point is not possible.

In the phase diagram, we can represent the experiment by a line between point (i) and point (f). (i) stands for "initial state," (f) for "final state." We move along an isotherm from (i) to (f).

2.14   What Happens When a Gas Is Compressed Below Its Critical Temperature?

Now we repeat the experiment below the critical point. We can compress water vapor at 200 °C (see Fig. 2.12) or butane at room temperature (C_4H_{10} has a critical point at about 152 °C).

Initially, compression proceeds similarly to the first experiment—both density and pressure increase. Reaching a certain volume, however, the particles are packed

Fig. 2.12 Isothermal compression of a gas below the critical point

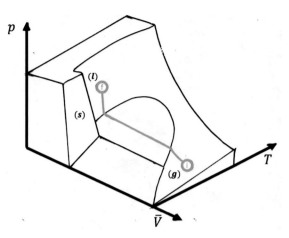

so tightly that their attractive forces cause the gas to condense and the first liquid droplets form.

After that, we can continue to compress, but the pressure remains constant, which means: we compress isobarically. Only when all gas has become liquid, then pressure will rise again.

In the phase diagram, this looks like that: Starting from (i), pressure increases, then the isotherm intersects the binodal [here: dew point curve], and then the system moves on a tie line through the two-phase region. On the tie line, both temperature and pressure are constant, only the volume decreases. Then we reach the intersection with the other binodal [here: boiling point curve] and finally the pressure increases very much until we reach the final state (f).

2.15   What Happens When Approaching the Critical Point?

Another experiment will illustrate the approach to the critical point:

We take an empty evacuated container with a volume of 56 mL and fill it with 18 g of water, i.e. 1 mol. We heat the system to 100 °C. We observe a pressure of 100 kPa and two phases: a denser phase, the liquid (density about 960 g/L) and a less dense phase, the gas phase (density about 0.6 g/L).

We further heat this two-phase system to 200 °C, thus increasing the pressure to 1.5 MPa and still retaining two phases. The liquid phase is now less dense than at 100 °C, while the gaseous phase is significantly denser than at 100 °C.

Gaseous and liquid phases thus approach each other, not only in terms of density, in terms of every property. We further increase the temperature and at 544 K (374 ° C) and 22 MPa the line of separation between liquid and gas disappears. We have reached the critical point. Figure 2.13 shows this isochoric process of approaching the critical point.

Fig. 2.13 Isochoric heating
of a two-phase system up to
the critical point

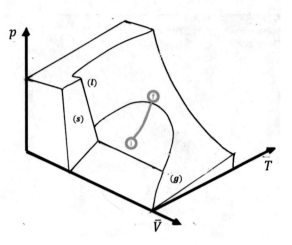

2.16 Summary

We should remember these equations and relationships:
The ideal gas law

$$pV = nRT \tag{2.25}$$

DALTON's law of partial pressure:

$$p_i = y_i \, p_{\text{total}} \tag{2.26}$$

$$p_{\text{total}} = \sum_i p_i \tag{2.27}$$

The *MAXWELL–BOLTZMANN*-theory with the velocity distribution and the average
velocity

$$\langle v \rangle = \sqrt{\frac{8RT}{\pi M}} \tag{2.28}$$

the average energy,

$$\langle U_{\text{trans}} \rangle = \frac{3}{2}R\,T \tag{2.29}$$

the mean free path

$$\langle \lambda \rangle = \frac{RT}{N_A\sqrt{2}\sigma p} \tag{2.30}$$

and last but not least the VAN-DER-WAALS equation,

$$\left(p + \frac{a}{\overline{V}^2}\right)\left(\overline{V} - b\right) = RT \tag{2.31}$$

Describing deviations from ideal behavior and explaining that at the critical point a gas is only 3/8 ideal.

$$p_{\text{c}}\,\overline{V}_{\text{c}} = \frac{3}{8}R\,T_{\text{c}} \tag{2.32}$$

2.17   Test Questions

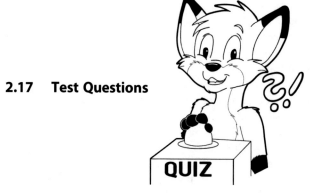

QUIZ

1. A balloon is filled with 1 L of air at 25 °C. The balloon is submerged 10 m under water. What is its volume now?
   (a) 1.0 L
   (b) 0.5 L
   (c) 0.25 L
   (d) 0.1 L

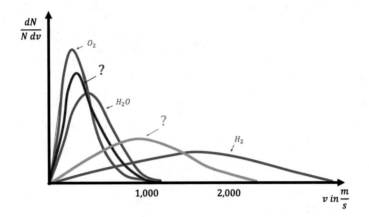

Fig. 2.14 MAXWELL–BOLTZMANN velocity distribution for several gases at 298 K

2. Air consists of 21 mol% oxygen (O_2) and 78 mol% nitrogen (N_2). 1 mol of air is present at standard conditions (0.0 °C, 100 kPa). Which statement(s) are correct?
   (a) Density = 1.3 g/L
   (b) Volume = 22.4 L
   (c) Thermal energy = 3.4 J
3. Dry air consists of 21 mol% oxygen (O_2), 78 mol% nitrogen (N_2) and about 1 mol% argon.
   1 mol of air is present at standard conditions (0.0 °C, 100 kPa).
      Which statement(s) are correct?
   (a) Ar and O_2 have the same average velocity
   (b) O_2 is (on average) faster than N_2
   (c) Ar and O_2 have the same average energy
   (d) Ar is slowest (on average)

4. Which statements are true for argon atoms in air at standard conditions?
   (a) average velocity ~ 1375 km/h
   (b) mean free path length ~ 0.075 μm
   (c) average surge frequency ~ 1000 MHz
   (d) mean translational energy ~ 0.35 eV (34 kJ/mol)
5. Which statements apply to a supercritical fluid?
   (a) The surface tension is zero
   (b) Strong compression results in the formation of two phases (liquid/gas)
   (c) The heat of evaporation is zero
   (d) The isotherm in the pV diagram shows kinks (=points where the curve is not differentiable)
6. Which gases belong to the "?" curves in Fig. 2.14?

   Left curve (red) corresponds to

- Helium (He)
- Neon (Ne)
- Argon (Ar)
- Nitrogen (N_2)
- Fluorine (F_2)
- Methane (CH_4)

Right curve (green) corresponds to

- Helium (He)
- Neon (Ne)
- Argon (Ar)
- Nitrogen (N_2)
- Fluorine (F_2)
- Methane (CH_4)

2.18 Exercises

1. In an auditorium ($V = 1000$ m^3), the relative humidity is 50.0%. The temperature is 20.0°C and the total pressure is 100 kPa. The vapor pressure of water at 20.0°C is 2.34 kPa (corresponding to 100% humidity or saturation).
   How much gaseous water is in the air of the auditorium?
2. The state of a gas was studied:
   Mass 6.00 g
   Volume 2.00 L
   Pressure 123 kPa
   Temperature 18.0 °C

Calculate the amount of substance of the gas and its molar volume and molar mass.

3. One mole of a gas with the molar mass $M = 83.0$ g/mol has the temperature $T = 83.0$ °C and the pressure $p = 101$ kPa.
   What is the average energy (thermal energy, translational energy) of the gas and how fast are the gas particles (average velocity)?

Thermal Equilibrium

3

3.1 Motivation

Spontaneous processes only run in the direction of equilibrium. But where is this equilibrium and how quickly can we reach it? And can we perhaps gain work on the way there, as a Stirling engine does, for example? (The *motivational picture* of this chapter, Fig. 3.1, illustrates heat conduction in a cylinder).

3.2 Where Does Equilibrium Lie and How Far Away Are We from Here?

Is our system in equilibrium? And if not, what is the distance from equilibrium?

Equilibrium is a very essential term in thermodynamics. Equilibrium is reached, when there is no more change in our system with time.

The easiest to describe are temperature and concentration equilibria. The system in Fig. 3.2 above left—consisting of two connected subsystems (1) and (2)—is obviously not in thermal equilibrium: the left subsystem has a higher temperature than the right subsystem.

Only after a certain time—in equilibrium—a uniform temperature is established (top right).

The equilibrium condition in this case is:

$$T_{1,eq} = T_{2,eq} \tag{3.1}$$

In all subsystems the temperature must be identical. The temperature difference ΔT between the subsystems in the left picture is a measure of the distance from equilibrium.

J. S. Lauth, *Physical Chemistry in a Nutshell*,
https://doi.org/10.1007/978-3-662-67637-0_3

Fig. 3.1 How quickly do we reach equilibrium and how much work can we gain in the process? (https://doi.org/10.5446/45977)

The system on the left in the center of Fig. 3.2 has a uniform temperature and yet is not in equilibrium, because the left subsystem shows a higher concentration than the right subsystem.

Equilibrium is defined here by the condition: "same concentration everywhere."

$$c_{1,eq} = c_{2,eq} \tag{3.2}$$

The concentration difference Δc in the initial state [two connected subsystems (1) and (2)] is a measure of the distance from equilibrium.

We can discuss chemical equilibria (Fig. 3.2 bottom row) quite similarly. The two "subsystems" (1) and (2) are the reactants and the products; the equilibrium condition is: "reactants and products show equal chemical potential"

$$\mu_{1,eq} = \mu_{2,eq} \tag{3.3}$$

We will discuss chemical equilibria in more detail in the next two chapters. In the following, we will discuss only physical (i.e., temperature and concentration) equilibria.

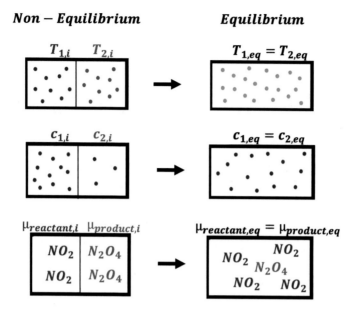

Non − Equilibrium   **Equilibrium**

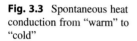

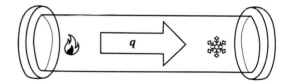

Fig. 3.2 Non-equilibrium and equilibrium (thermal equilibrium, concentration equilibrium, chemical equilibrium)

Fig. 3.3 Spontaneous heat conduction from "warm" to "cold"

3.3   How Fast Does a System Go into Equilibrium by Conduction?

If there are different temperatures in a system (Fig. 3.3), this provokes heat transport. It is a passive energy transport without external flow−this is referred to as heat conduction.

If different concentrations are present in a system (Fig. 3.4), mass transport (or mass transfer) is provoked; this is referred to as passive mass transfer or diffusion.

Both heat conduction and diffusion are passive; they are summarized under the term conduction.

Active transport processes associated with a flow are called convection. Flow processes (a distinction is made between laminar and turbulent flow, for example) are extremely important in nature and technology and are therefore discussed in detail in special courses on fluid mechanics. Here we want to concentrate only on

Fig. 3.4 Spontaneous
diffusion from "concentrated"
to "diluted"

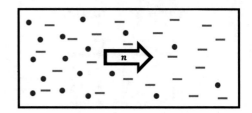

Fig. 3.5 Temperature profile
for non-stationary heat
conduction

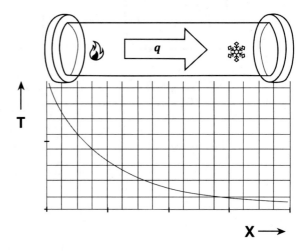

passive transport processes. We exclude flow completely, we are only dealing with
fluids at rest.

3.4   How Fast Does Heat Flow by Conduction?

Heat conduction can be described quantitatively by *FOURIER*'s laws.

A prerequisite for heat conduction is a temperature gradient in the system
(Fig. 3.5).

In the one-dimensional case, this gradient will be described by the first derivative
of the temperature profile

$$\frac{dT}{dx} \tag{3.4}$$

Gradient of the temperature profile

The passively transported amount of heat is quantified by the flux density—the
amount of heat that flows through a surface per second

$$\frac{dq}{A\,dt} \tag{3.5}$$

Heat flux density

FOURIER'S 1ST law of heat conduction describes mathematically that the flux density is proportional to the gradient.

$$\frac{dq}{A\,dt} = -\lambda\,\frac{dT}{dx} \tag{3.6}$$

λ is the thermal conductivity coefficient and is a measure of how well or poorly a medium at rest conducts heat. Metals have a very high thermal conductivity; quiescent liquids and especially quiescent gases conduct heat very poorly.

The steeper the gradient, the greater the heat flow. In Fig. 3.5, the gradient on the left is the steepest—this is where the most heat flows.

If the gradient is zero, the heat flow is also zero (thermal equilibrium). In Fig. 3.5, the heat flux is not the same everywhere; this is referred to as non-stationary heat conduction. In this case, the temperature profile will change over time.

3.5   How Does the Temperature Profile Change?

In the case of non-stationary heat conduction, the temperature profile exhibits a curvature that can be described by the second derivative of the temperature profile.

$$\frac{d^2T}{d\,x^2} \tag{3.7}$$

Curvature of the temperature profile

In that case, temperature changes occur in the system, which are described by *FOURIER'S 2ND* law of heat conduction.

$$\frac{dT}{dt} = \frac{\lambda}{\varrho\,c_p}\,\frac{d^2T}{d\,x^2} \tag{3.8}$$

In Fig. 3.5, the curvature of the profile is greatest at about the center—this is where the temperature changes most significantly with time. At the edge of the temperature profile in Fig. 3.5, the curvature is almost zero—here the temperature remains essentially constant.

3.6   How Well Does a (Quiescent) Gas Conduct Heat?

In gases at rest, heat transport occurs via collisions. The thermal conductivity of gases at rest can be explained by the kinetic theory of gases.

Gas (20 °C, 100 kPa)	Thermal conductivity λ
Xenon (Xe)	$\lambda = 0.005 \frac{W}{K\,m}$
Air	$\lambda = 0.03 \frac{W}{K\,m}$
Hydrogen (H$_2$)	$\lambda = 0.18 \frac{W}{K\,m}$

Table 3.1 Thermal conductivity of some (quiescent) gases

$$\lambda = \frac{25\,\pi}{64}\, \overline{C_V}\, \langle \lambda \rangle\, \langle v \rangle\, \frac{n}{V} \qquad (3.9)$$

The thermal conductivity of a gas is related to its average velocity $\langle v \rangle$ and mean free path $\langle \lambda \rangle$. Small and light gas molecules therefore have the highest thermal conductivity (see Table 3.1).

A simple light bulb can be regarded as a thermal conductivity detector (TCD): the worse the thermal conductivity of the filling gas, the brighter the filament lights up. This phenomenon is used to detect gases, e.g. in gas chromatographs.

Gases are extremely poor conductors of heat—but it should be remembered once again that we have completely excluded flow from our considerations. In practice, gases conduct heat much better due to convection phenomena. However, if we prevent convection in gases (for example, by foaming like in Styrofoam®), gases actually insulate thermally very well.

3.7 How Fast Does Diffusion Proceed?

The quantitative description of diffusion is given by FICK'S laws. These describe the flux density and the concentration change completely analogous to FOURIER'S laws as a function of the slope and curvature of a profile. The cause for diffusion is a concentration gradient, in the one-dimensional case the slope of the concentration profile.

$$\frac{dc}{dx} \qquad (3.10)$$

Gradient of the concentration profile

The amount of mass transported passively (without flow) is quantified by the flux density.

$$\frac{dn}{A\,dt} \qquad (3.11)$$

Flux density of the amount of substance

FICK'S 1ST law of diffusion describes mathematically that the flux density is proportional to the gradient.

Table 3.2 Diffusion constants in fluid, gaseous, and solid media	Component/medium	Diffusion constant D
	Sugar/water (20 °C)	$D = 5 \cdot 10^{-10} \frac{m^2}{s}$
	Carbon dioxide/air (20 °C, 100 kPa)	$D = 2 \cdot 10^{-5} \frac{m^2}{s}$
	Carbon/iron (800 °C)	$D = 2 \cdot 10^{-13} \frac{m^2}{s}$

Fig. 3.6 Concentration profile with non-stationary diffusion

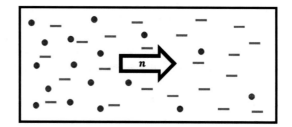

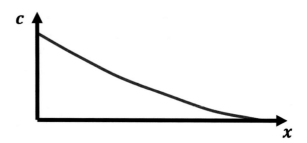

$$\frac{dn}{A \, dt} = -D \frac{dc}{dx} \tag{3.12}$$

The coefficient D describes the diffusion in a medium at rest. This coefficient is naturally much smaller with diffusion in a liquid than diffusion in a gas (see Table 3.2).

Where the concentration profile has the largest negative slope (far left in Fig. 3.6) diffusion will be fastest from left to right.

3.8   How Does the Concentration Profile Change?

In the case of non-stationary diffusion, the concentration profile exhibits a curvature that can be described by the second derivative of the temperature profile.

$$\frac{d^2 c}{d x^2} \tag{3.13}$$

Curvature of the concentration profile

In non-stationary diffusion, the concentration in the system will change with time. The change in concentration is proportional to the curvature of the profile. This is the *FICK'S 2ND* law of diffusion.

$$\frac{dc}{dt} = D\,\frac{d^2c}{d\,x^2} \tag{3.14}$$

In Fig. 3.6, the curvature of the profile is greatest at about the center—this is where the concentration will change most significantly with time.

3.9 How Fast Do Gases Diffuse?

The diffusion of gases can also be explained by the kinetic theory of gases:

$$D \sim \frac{3\pi}{16}\,\langle v \rangle\,\langle \lambda \rangle \tag{3.15}$$

Small and light gas particles diffuse fastest. *EINSTEIN* and *SMOLUCHOWSKI* could explain *FICK'S* laws using their "random walk" model (see Fig. 3.7).

In particular, they were able to calculate the displacement x—which is an indication of how far a particle moves from its starting point by diffusion.

$$\langle x2 \rangle = 2 \cdot D \cdot t \tag{3.16}$$

3.10 How Do Energy and Entropy Change During Heat Transport?

Let us discuss a temperature equilibrium experiment thermodynamically (see Fig. 3.8).

Fig. 3.7 Random walk of a diffusing particle

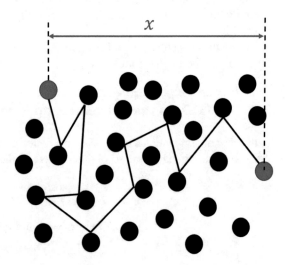

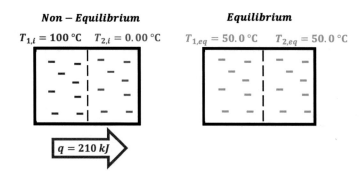

Fig. 3.8 Experiment for establishing thermal equilibrium

Our two subsystems (1) and (2) at the beginning of the experiment are:

(1) 1.00 kg of water at 100 °C and
(2) 1.00 kg of water at 0.00 °C.

After a certain time, thermal equilibrium has been reached: both systems now show a temperature of 50.0 °C.

We can apply the basic equation of calorimetry and calculate the exchanged heat q. The heat capacities of both systems are equal.

$$C_p(1) = C_p(2) = 4.184 \frac{kJ}{K} \tag{3.17}$$

the temperature difference has the same amount in both cases

$$\Delta T(2) = -\Delta T(1) = 50\,°C = 50\ K \tag{3.18}$$

(in case of temperature differences, we may use °C or K without conversion)
We calculate the amount of heat transferred from system 1 to system 2.

$$q(2) = C_p(2)\Delta T(2) = 4.184 \frac{kJ}{K}\ (50\ K) = 209\ kJ \tag{3.19}$$

We now want to continue in the tradition of thermodynamics and balance both energy and entropy of this process using the first and second law.

3.11   What Is Internal Energy and What Does the First Law of Thermodynamics State?

The letter U stands for the thermodynamic quantity of Internal Energy. This is a measure of how much energy is contained in a system. The introduction of this quantity is meaningful, because the first law states that the total energy of the universe remains constant.

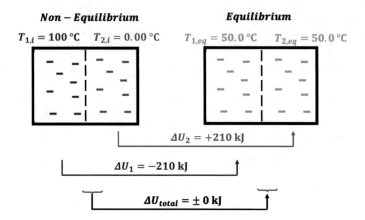

Fig. 3.9 Energy balance when approaching thermal equilibrium

$$\Delta U + \Delta U_{\text{sur}} = 0 \tag{3.20}$$

We can measure the change in Internal Energy of a process by simply adding up exchanged heat and work.

$$\Delta U = q + w \tag{3.21}$$

So, in any process where heat or work is involved, the Internal Energy changes.

In particular, these are temperature change, phase change, and chemical reactions. Note however, that when a (ideal) system is diluted, the energy does not change (5 L of air has the same energy as 4 L of nitrogen and 1 L of oxygen).

The elements at 25 °C were arbitrarily chosen as the zero level for energy. Most compounds have lower energy than the elements and therefore have a negative value for U.

Let us look at our example process through "energy-tinted spectacles" and apply the first law: The originally hot water [subsystem (1)] has decreased its Internal Energy by 209 kJ. The originally cold water [subsystem (2)] has increased its internal energy by 209 kJ (see Fig. 3.9).

3.12   What Is Entropy and What Does the Second Law of Thermodynamics State?

The letter S stands for the thermodynamic quantity of entropy. The entropy is a measure of the chaos in a system. The introduction of this quantity is meaningful, because the second law of thermodynamics states that the total entropy in the universe can only increase.

Fig. 3.10 Mnemonic to memorize CLAUSIUS' definition of entropy (q-T for Cutie; in German "Kuh durch Tee")

$$\Delta S + \Delta S_{\text{sur}} \geq 0 \tag{3.22}$$

According to CLAUSIUS, change of entropy during a process can be measured by the so-called reduced heat.

$$\Delta S = \frac{q_{\text{rev}}}{T} \tag{3.23}$$

This little mnemonic can help us remember CLAUSIUS' equation (see Fig. 3.10).

Entropy does show similar dependencies as Internal Energy, i.e. it changes whenever heat occurs in a process (temperature change, phase change, chemical reaction). In addition, however, it is also dependent on dilution. 5 L of air have a higher entropy than 4 L of nitrogen and 1 L of oxygen.

The zero point of entropy is an ideal crystal at 0 Kelvin. This is the third law of thermodynamics. All elements and compounds at room temperature therefore have positive values for S at room temperature.

We look at our example process through "entropy-tinted spectacles": The entropy of the originally hot water [subsystem (1)] has decreased, heat has been released (convention: negative sign = energy release).

To calculate the numerical value of entropy change, we have to integrate CLAUSIUS' formula, because temperature is not constant during the process.

$$\Delta S = \frac{q_{\text{rev}}}{T} \tag{3.24}$$

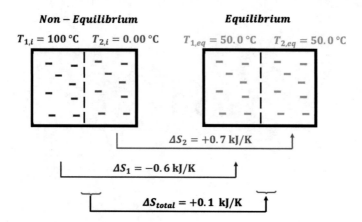

Fig. 3.11 Entropy balance when approaching thermal equilibrium

$$\Delta_{T_i \to T_f} S = \int \frac{q_{rev}}{T} = C \ln \frac{T_f}{T_i} \tag{3.25}$$

$$\Delta_{T_i \to T_f} S = 4.184 \frac{kJ}{K} \ \ln \left(\frac{323 \ K}{373 \ K} \right) = -0.60 \frac{kJ}{K} \tag{3.26}$$

The entropy of subsystem (1) has decreased by 0.60 kJ/K. Using the same formula, we can calculate that the entropy of the originally cold water has increased—but by a larger amount (see Fig. 3.11).

All in all, entropy was generated during the process. According to the second law, this process will never spontaneously run in the reverse direction.

Processes in which the total entropy increases are called irreversible. They can only take place in ONE direction. A spontaneous heat transport from hot to cold is irreversible, as just described.

3.13   How Much Heat Can We Convert into Work?

CARNOT thought about how to extract work from the just discussed process. He described an engine that reverses the flow of heat from hot to cold and thus extracts the maximum possible amount of work from heat flow. The CARNOT engine is thus an ideal heat engine; it converts heat into work with the best possible efficiency η_C (see Fig. 3.12).

A Stirling engine is also a heat engine: it also consists of two temperature levels between which heat flows and a certain fraction of which is converted into work.

The efficiency of this real heat engine must be lower than the efficiency of the ideal CARNOT engine.

By the way, both engines work with a gas as working medium; especially in the CARNOT process, this gas runs through a cycle of two isotherms and two adiabats.

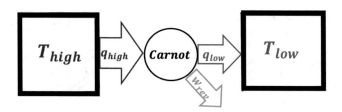

Fig. 3.12 CARNOT engine working as a heat engine

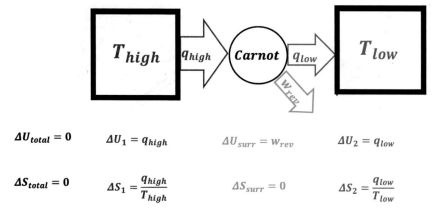

$\Delta U_{total} = 0$ $\qquad \Delta U_1 = q_{high}$ $\qquad \Delta U_{surr} = w_{rev}$ $\qquad \Delta U_2 = q_{low}$

$\Delta S_{total} = 0$ $\qquad \Delta S_1 = \dfrac{q_{high}}{T_{high}}$ $\qquad \Delta S_{surr} = 0$ $\qquad \Delta S_2 = \dfrac{q_{low}}{T_{low}}$

Fig. 3.13 Energy and entropy balance of a CARNOT engine

However, these details of the mode of operation are not important for the following energy and entropy balance.

Let u discuss a complete working cycle of a CARNOT engine balancing both energy and entropy ("using energy and entropy-tinted spectacles").

Let us consider step by step the four subsystems (subsystem (1), subsystem (2), CARNOT engine, surroundings) with respect to their energy and entropy.

When the CARNOT engine is running, both internal energy and entropy of the upper temperature level (subsystem 1) decrease. For the lower temperature level (subsystem 2), the reverse is true: both internal energy and entropy increase.

The CARNOT engine itself goes through a circular process and returns to its initial state; thus, in sum, it changes neither energy nor entropy.

The surroundings absorb work; thus, they change their internal energy. The entropy of the surroundings, however, remains constant (see Fig. 3.13).

3.14   How Do We Calculate the Efficiency of a CARNOT Engine Using the Laws of Thermodynamics?

According to the first law, the exchanged energies must add up to 0:

$$q_{high} + q_{low} + w_{rev} = 0 \tag{3.27}$$

Energy can neither be created nor destroyed.

According to the second law, the total entropy can never decrease, which means that the sum of the entropy amounts must be greater than or equal to 0.

$$\frac{q_{high}}{T_{high}} + \frac{q_{low}}{T_{low}} = 0 \tag{3.28}$$

The CARNOT engine works ideally reversible, which means the total entropy change is equal to 0.

By combining these two equations, we obtain the CARNOT efficiency to be

$$\eta_C = \frac{-w_{rev}}{q_{high}} \tag{3.29}$$

$$\eta_{Carnot} = \frac{T_{high} - T_{low}}{T_{high}} \tag{3.30}$$

This is one of the most important equations of thermodynamics. It limits the conversion of heat into work.

The temperature difference of the levels in a heat engine determines the efficiency. An ideal heat engine with the two temperature levels 0 °C and 100 °C therefore has an efficiency of

$$\eta_{Carnot} = \frac{T_{high} - T_{low}}{T_{high}} = \frac{373\ K - 273\ K}{373\ K} = 0.27 \tag{3.31}$$

i.e., approx. 27%: Out of 100% heat from the high temperature level, only 27% is converted to work, the remaining 75% flows as "waste heat" into the low temperature level.

3.15 Summary

The conductive transport processes thermal conductivity and diffusion can be described by similar laws: these are *FOURIER'S* laws

$$\frac{dq}{A\,dt} = -\lambda\,\frac{dT}{dx} \qquad \frac{dT}{dt} = \frac{\lambda}{\varrho\,c_p}\,\frac{d^2T}{d\,x^2} \tag{3.32}$$

and *FICK'S* laws.

$$\frac{dn}{A\,dt} = -D\,\frac{dc}{dx} \qquad \frac{dc}{dt} = D\,\frac{d^2c}{d\,x^2} \tag{3.33}$$

Small and light gas particles transport heat best and also diffuse fastest.

$$\lambda = \frac{25\,\pi}{64}\,\overline{C_V}\,\langle\lambda\rangle\,\langle v\rangle\,\frac{n}{V} \qquad D = \frac{3\,\pi}{16}\,\langle\lambda\rangle\,\langle v\rangle \tag{3.34}$$

While the complete conversion of work into heat is possible without problems, the conversion of heat into work is possible only to a certain extent; according to *CARNOT*, this efficiency cannot exceed

$$\eta_{\text{Carnot}} = \frac{T_{\text{high}} - T_{\text{low}}}{T_{\text{high}}} \tag{3.35}$$

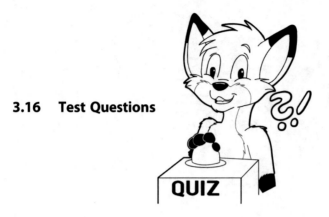

3.16   Test Questions

1. Mark the correct statement(s)
   (a) Ammonia (NH_3) diffuses in air faster than hydrogen chloride (HCl)
   (b) Hydrogen diffuses better at 50 °C than at 25 °C
   (c) In steady-state diffusion, the mass flow density is the same everywhere
   (d) (Quiescent) Argon conducts heat better than helium at standard conditions
       (STP, 0 °C, 100 kPa)
       (both gases have identical heat capacities $\overline{C_V}$)

2. Which statements are true for CARNOT'S cycle?
   (a) Entropy change at T_{high} = −Entropy change at T_{low}
   (b) Energy change at T_{high} = −Energy change at T_{low}
   (c) The process runs with the same efficiency in both directions (as heat pump
       and heat engine)
   (d) The efficiency can never be 100%

3. In a medium at rest, the following concentration profile exists at a time t (see
   Fig. 3.14):

Fig. 3.14 Concentration
profile in a cuvette

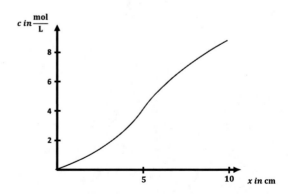

(a) Where is the mass flow density $\frac{dn}{A \cdot dt}$ greatest?

(b) Where is the change in concentration over time $\frac{dc}{dt}$ greatest?

3.17 Exercises

1. A window with dimensions 1.00 m × 1.00 m consists of a single pane of glass with a thickness of 4.00 mm. On the outside, the temperature is 18.0 °C; on the inside, 20.0 °C. How much heat q is transported through the glass pane per second?

Thermal conductivity of glass: 0.760 W/(°C m)

2. A CARNOT heat pump takes in a quantity of heat q_{low} at 0.00 °C, transports ("pumps") it to a higher temperature level (25.0 °C) with the aid of work w_{rev}, and releases a quantity of heat q_{high} at this temperature level.

How much work w_{rev} does the engine need to output $q_{high} = -500$ kJ at the higher temperature?

What is the efficiency η of the engine ($\eta = -w_{rev}/q_{high}$)?

3. A coal-fired power plant operating between 500 °C (superheated steam) and 100 °C (condenser) and having 80.0% of the theoretically possible (reversible) efficiency supplies 50.0 MJ of electrical work per second. Create the energy balance of this power plant:

How much heat is absorbed per second $\dot{q}_{high}$?

How much waste heat is emitted per second $\dot{q}_{low}$?

Affinity

4

4.1 Motivation

Some processes run voluntarily, so they do have an impetus (or affinity), other processes do not. We will see in this chapter how we can use the First and Second Law to predict the impetus of a process. (The *motivational picture* of this chapter, Fig. 4.1, illustrates the change of free enthalpy with evaporation of water).

We will see in this chapter how we can use the First and Second Law to predict the impetus of a process.

4.2 How Much Internal Energy Is in a System?

At the end of the nineteenth century, the opinion prevailed that the impetus of a process is related to the change of energy. *Berthelot's* principle demands that only exothermic processes can take place spontaneously.

Today we know that in addition to energy change, entropy change plays a crucial role in affinity.

So the question about impetus can also be formulated as: "Does the energy decrease during the process and/or does the entropy increase during the process?" Or more boldly, "Are energy and/or entropy with us?"

We consider the autoprotolysis of water as an example—the splitting of water into H^+ and OH^- ions.

$$H_2O(l) \rightarrow H^+(aq) + OH^-(aq) \tag{4.1}$$

Does this process have an impetus? What about energy change and entropy change in this process?

J. S. Lauth, *Physical Chemistry in a Nutshell*, https://doi.org/10.1007/978-3-662-67637-0_4

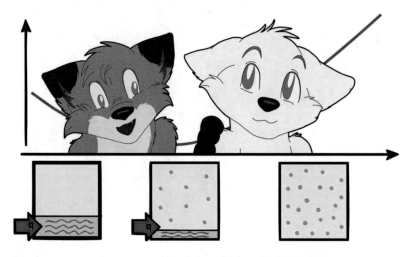

Fig. 4.1 Are energy and/or entropy with us? (https://doi.org/10.5446/45978)

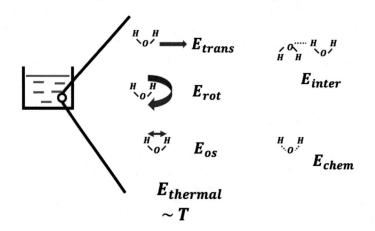

Fig. 4.2 Internal energy in water

Thermodynamics knows the quantity "internal energy" U, which summarizes all energy in a system (see Fig. 4.2).

Where is internal energy located in water, for example? Well, the water molecules are not at rest, but they have kinetic energy, more precisely: translational, rotational, and oscillatory energy. There are also energies between different molecules and, very importantly, chemical bonding energies between atoms. All these energies (and more) add up to internal energy.

$$U = \sum E = E_{\mathrm{trans}} + E_{\mathrm{rot}} + E_{\mathrm{os}} + E_{chem} + \dots \tag{4.2}$$

$$E_{\mathrm{trans}} + E_{\mathrm{rot}} + E_{\mathrm{os}} = E_{\mathrm{therm}} \tag{4.3}$$

4.3   How Can We Measure Internal Energy?

We cannot measure absolute values of internal energy, but its change can be measured.

This does not make the quantity U less valuable: e.g. we cannot measure the absolute height of an object either, only relative heights to an arbitrary zero level can be measured. However, changes in height can be determined independently of the zero level.

Whenever we add or remove energy to or from a system, its internal energy changes. Supply and removal can take place as heat q or as work w.

$$\Delta U = q + w \qquad (4.4)$$

So if we want to calculate ΔU for the oxyhydrogen reaction, we simply have to add the heat and work amounts of this reaction. For the reversible path, these are the reversible heat q_{rev}, the pressure–volume work w_{pV}, and the electric work w_{el}.

$$\Delta U = (-98) + (-474) + (+7.4) = -564.6 \text{ kJ} \qquad (4.5)$$

The product water has 565 kJ less energy than the reactants hydrogen and oxygen.

The equation for the calculation of ΔU when there is no work involved in the process ($w = 0$), and this is the case when the volume remains constant in a spontaneous process. Constant volume means: no pressure–volume work.

$$w_{pV} = -p\Delta V \qquad (4.6)$$

The heat q_V released during an isochoric (spontaneous) process is identical to the change of internal energy ΔU.

$$\text{if} \quad w = 0 \qquad (4.7)$$

$$\Delta U = q_V \qquad (4.8)$$

4.4   How Do We Turn Isobaric Heat into a State Variable?

If, on the other hand, volume does change in isobaric processes, we must always consider pressure–volume work as well.

$$w_{pV} = -p\Delta V \qquad (4.9)$$

This means that with isobaric spontaneous processes the heat we measure—the isobaric heat q_p—corresponds to a sum of two energies.

$$q_p = \Delta U + p\Delta V \tag{4.10}$$

Because we very often measure isobaric heats, the combination of state variables $(U + p\,V)$ was simply renamed to "enthalpy." The enthalpy is an "artificial quantity" with no easily interpretable graphic image, but the change of enthalpy is always equal to the (spontaneous) isobaric heat.

$$q_p = \Delta U + p\Delta V \equiv \Delta H \tag{4.11}$$

In general, physical quantities do not have to be easily interpretable—serious science only requires that every quantity is useful, on the one hand, and must be measurable on the other hand.

In the oxyhydrogen reaction, we were able to measure an isobaric heat of − 572 kJ—this also corresponds to the reaction enthalpy.

$$\Delta H = q_p = -572 \text{ kJ} \tag{4.12}$$

The product water is 572 kJ lower in enthalpy than the reactants hydrogen and oxygen.

So we may complete enthalpy's profile.

4.5   What Is Enthalpy? (Factsheet: Table 4.1)

Enthalpy is a measure of energy content; useful for applicating the first law. Enthalpy change can be measured simply by measuring isobaric heat q_p.

Whenever isobaric heat occurs, there is also an enthalpy change in the system. Consequently, enthalpy depends on temperature, on phase, and on chemical

Table 4.1 Enthalpy factsheet

Enthalpy H	
H is a measure of	Energy in the system
Statement of the First Law	The energy of the universe is constant
Measurement of ΔH	Measurement as isobaric heat $\Delta H = q_p$
Calculation of ΔH	$\Delta H = H_{final} - H_{initial}$
H depends on	• Chemical structure • Temperature • Phase
Zero point of H	Elements in standard state, 100 kPa, 25 ° C
Table values	Standard enthalpies of formation $\Delta_f H°$
Example	$\Delta_f H°\,(H_2) = 0\,\frac{kJ}{mol}$ $\Delta_f H°\,(O_2) = 0\,\frac{kJ}{mol}$ $\Delta_f H°\,(H_2O(l)) = -285.84\,\frac{kJ}{mol}$ $\Delta_f H°\,(H_2O(g)) = -241.83\,\frac{kJ}{mol}$

structure. The zero point of enthalpy has been arbitrarily set: All elements in the most stable form have enthalpy 0 at 25 °C. Enthalpies that include this zero point are called standard enthalpy of formation $\Delta_f H°$.

Most compounds have negative standard enthalpies of formation. This means: these compounds are lower in energy than the elements.

The enthalpy change in a process can always be calculated as

$$\Delta_{initial \to final} H° = \Delta_f H° \text{ (final state)} - \Delta_f H° \text{ (initial state)} \tag{4.13}$$

4.6   When Does Enthalpy Change?

The enthalpy changes whenever isobaric heat is absorbed or released, i.e. when temperature changes

$$\Delta_{T_1 \to T_2} H = q_{p,T_1 \to T_2} = \int_{T_1}^{T_2} C_p \, dT \tag{4.14}$$

with a phase change

$$\Delta_{s \to l} H = q_{p,\text{fus}} \tag{4.15}$$

or in a chemical reaction.

$$\Delta_r H = q_{p,\text{rxn}} \tag{4.16}$$

$$\Delta_r H° = \Delta_f H° \text{ (Products)} - \Delta_f H° \text{ (Reactants)} \tag{4.17}$$

On the other hand, if we mix two ideal systems, no heat is released or absorbed: there is no enthalpy of mixing in ideal systems.

$$\Delta_{mix} H = q_{mix} = 0 \text{ (ideal)} \tag{4.18}$$

4.7   Does Enthalpy Change Depend on the Path?

If we want to convert 1 mole of ice at 0 °C into 1 mole of water at 0 °C, we have to add 6 kJ of heat to do so. If we evaporate the liquid water afterwards at 0 °C, we need another 45 kJ of heat. We could also evaporate the ice immediately (sublimation), in which case the heat required is equal to the sum of these two amounts, i.e. 51 kJ (see Fig. 4.3).

The isobaric heat—or, as we say now—the enthalpy change ΔH is therefore (like any other change of state variable ΔZ) not path-dependent. This statement is the so-called HESS's Law: The enthalpies of reaction are path-independent.

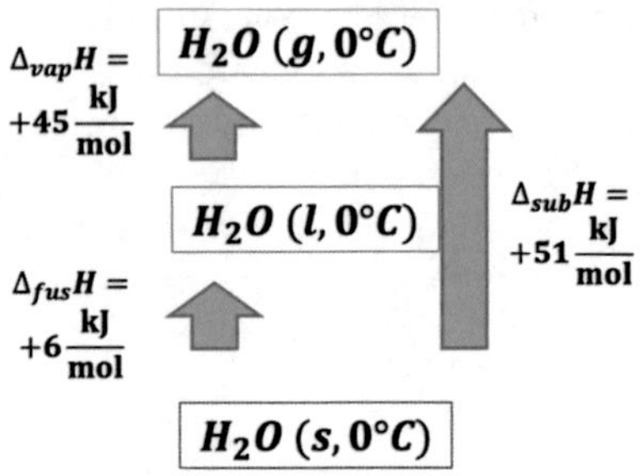

Fig. 4.3 Phase transformation enthalpies of water at 0 °C to illustrate HESS's Law

$\Delta_{vap}H = +45 \dfrac{kJ}{mol}$

$H_2O\ (g, 0°C)$

$H_2O\ (l, 0°C)$

$\Delta_{sub}H = +51 \dfrac{kJ}{mol}$

$\Delta_{fus}H = +6 \dfrac{kJ}{mol}$

$H_2O\ (s, 0°C)$

Fig. 4.4 Enthalpy diagram for the enthalpy of formation and atomization of water

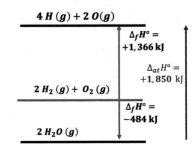

$4\,H\,(g) + 2\,O(g)$

$\Delta_f H° = +1,366\ kJ$

$\Delta_{at} H° = +1,850\ kJ$

$2\,H_2\,(g) + O_2\,(g)$

$\Delta_f H° = -484\ kJ$

$2\,H_2O\,(g)$

$$\Delta_{A \to C}H = \Delta_{A \to B}H + \Delta_{B \to C}H \tag{4.19}$$

HESS's Law is a special form of the law of conservation of energy; however, it was formulated before the first law.

4.8 How Much Enthalpy Is Present in a Molecule or in a Chemical Bond?

Let us draw water and its components in an enthalpy diagram. The elements $H_2(g)$ and $O_2(g)$ mark the zero level. The atoms $H(g)$ and $O(g)$ mark a much more energetic state than the elements and water is energetically much lower than the elements.

The arrows in Fig. 4.4 starting from the zero level are enthalpies of formation: the down arrow is the enthalpy of formation of water; the up arrow symbolizes the enthalpies of formation of the atoms. According to HESS' theorem, we can add these arrows vectorially:

The arrow connecting the water to the atoms is the atomization enthalpy $\Delta_{at}H°$ of water: the heat we need to split water molecules completely into atoms. We can obtain this heat according to the theorem of HESS by vectorial combination of the enthalpies of formation

$$\Delta_{at}H° = (4\ \Delta_fH\ (\text{H}) + 2\ \Delta_fH\ (\text{O})) - (2\ \Delta_fH\ (\text{H}_2\text{O})) \qquad (4.20)$$

$$\Delta_{at}H° = (1366\ \text{kJ}) - (-484\ \text{kJ}) = 1850\ \text{kJ} \qquad (4.21)$$

Conversely, these 1850 kJ would be released again when we form two water molecules from the atoms. This means: -1850 kJ corresponds to four times the bond enthalpy of the OH bond.

$$\Delta_{at}H° \approx -\sum \langle H_{bond}\rangle \qquad (4.22)$$

Bond enthalpies are usually average values; therefore the equation is only approximately correct.

4.9   How Does Enthalpy Change During a Reaction?

With the tabulated enthalpies of formation, we can calculate arbitrary heats of reaction or enthalpies of reaction $\Delta_rH°$ can be calculated. For example, if we want to know how much heat is released or consumed when water is split into H^+ and OH^-, we draw an enthalpy diagram (see Fig. 4.5):

Again we draw the elements as zero levels as well as the reactant (water) and the products (H^+ and OH^-) according to their enthalpies of formation.

The arrows starting from the zero level are, as we already know, the enthalpies of formation; the arrow between reactant and products is the enthalpy of reaction

$$\Delta_rH° = \Delta_fH°\,(\text{Products}) - \Delta_fH°\,(\text{Reactants}) \qquad (4.23)$$

By vectorial combination of the enthalpies of formation, we obtain the enthalpy of reaction. We have to subtract two negative numbers from each other in the calculation and get the reaction enthalpy ("stoichiometric sum of enthalpies").

$$\Delta_rH° = (\Delta_fH°\,(\text{H}^+) + \Delta_fH°\,(\text{OH}^-)) - (\Delta_fH°\,(\text{H}_2\text{O})) \qquad (4.24)$$

Fig. 4.5 Enthalpy diagram for autoprotolysis reaction

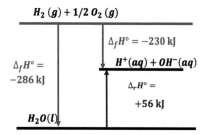

$$\Delta_r H^\circ = (-230 \text{ kJ}) - (-286 \text{ kJ}) \qquad (4.25)$$

$$\Delta_r H^\circ = -56 \text{ kJ}. \qquad (4.26)$$

This means that the autoprotolysis of water is endothermic: we would have to invest 56 kJ of heat to dissociate 1 mole of water completely to 1 mole of H^+ and 1 mole of OH^-. The process is energetically "uphill"; so energy is obviously "not with us."

The enthalpies of formation are given in the tables as molar quantities in the unit kJ/mol.

When we calculate enthalpy change in a chemical reaction, we may need to consider the stoichiometric numbers.

For example, if we formulate the oxyhydrogen reaction as

$$2\,H_2(g) + O_2(g) \rightarrow 2\,H_2O(l) \qquad (4.27)$$

we obtain a reaction enthalpy of $-572\,\frac{kJ}{mol}$

"mol" in this context means "formula conversion." The reaction equation should always be included when giving $\Delta_r H^\circ$.

The enthalpy of reaction can also be estimated from the bond enthalpies if only gaseous substances are involved in the reaction.

$$\Delta_r H^\circ \approx \sum \langle H_{bond}\rangle (\text{Products}) - \sum \langle H_{bond}\rangle (\text{Reactants}) \qquad (4.28)$$

4.10   What Is Entropy? (Factsheet: Table 4.2)

From entropy's profile, we see that it is a measure of chaos. It makes sense to introduce this quantity, because the second law can be easily formulated using entropy. We can measure entropy changes according to CLAUSIUS as reduced heat (quotient of reversible heat and temperature). Rule of thumb for this: "Cutie [q/T]."

Entropy depends on temperature, phase, and chemical structure just like energy, but additionally on dilution. Even in ideal systems, entropy increases with dilution.

4.11   How Can We Measure Entropy?

Entropy has an absolute zero point—this is stated by the third law—pure crystals at 0 K do not possess any entropy.

As a result, practically only positive numbers can be found in the table values for entropy. The measurement of entropy according to CLAUSIUS is done via the reduced reversible heat.

Table 4.2 Entropy factsheet

Entropy S	
S is a measure of	Chaos (Disorder, Neginformation)
Statement of the Second Law	The entropy of the universe can only increase (irreversible processes) *or stay the same (reversible processes).*
Measurement of ΔS	Measurement according to Clausius as reduced reversible heat $\Delta S = \frac{q_{rev}}{T}$ Calculation according to Boltzmann from thermodynamic probabilities $S = k \ln (\Omega)$
Calculation of ΔS	$\Delta S = S_{final} - S_{initial}$
S depends on	• chemical structure • Temperature • Phase • *Dilution*
Zero point of S	Ideal crystals at 0 Kelvin (third Law)
Table values	Standard entropy $S°$
Examples	$S°(H_2) = 130.684 \frac{J}{mol\ K}$ $S°(O_2) = 205.0 \frac{J}{mol\ K} S°(H_2O(l)) = 69.9 \frac{J}{mol\ K}$ $S°(H_2O(g)) = 188.72 \frac{J}{mol\ K}$

$$\Delta S = \frac{q_{rev}}{T} \tag{4.29}$$

The index "reversible" is sometimes very important: if we run the oxyhydrogen reaction spontaneously, we measure a lot of heat that is released—but this is not the reversible heat we need. We have to run the oxyhydrogen reaction reversibly over a fuel cell, then we get −98 kJ as reversible heat.

Only this heat we may insert into CLAUSIUS' formula.

$$\Delta S = \frac{q_{rev}}{T} = \frac{-98\ kJ}{298\ K} = -326 \frac{J}{K} \tag{4.30}$$

Entropy can also be calculated according to BOLTZMANN'S statistical theory.

$$S = k \ln (\Omega) \tag{4.31}$$

(famous formula, engraved in Boltzmann's tombstone)

Ω is the so-called thermodynamic probability. But we do not use this definition in our course (for more information: see textbooks of statistical thermodynamics).

4.12   When Does Entropy Change?

So entropy changes whenever we have to add or remove reversible heat: During temperature changes, during phase changes, and during chemical reactions. In addition, entropy increases with mixing.

Entropy change with temperature change:

$$\Delta_{T_1 \rightarrow T_2} S = \int \frac{q_{p,T_1 \rightarrow T_2}}{T} = \int_{T_1}^{T_2} \frac{C_p}{T} \, dT \tag{4.32}$$

Entropy change with phase change:

$$\Delta_{s \rightarrow l} S = \frac{q_{rev,fus}}{T_{fus}} \tag{4.33}$$

$$\Delta_r S = \frac{q_{rev,rxn}}{T} \tag{4.34}$$

Entropy change during mixing (volume increase):

$$\Delta_{mix} S = \sum nR \ln \frac{V_f}{V_i} \tag{4.35}$$

4.13   How Does Entropy Change in a Reaction?

We calculate reaction entropies in the same way as we calculate reaction enthalpies from the table values with the aid of an equation analog to Hess' law.

$$\Delta_r S^\circ = S^\circ (\text{Products}) - S^\circ (\text{Reactants}) \tag{4.36}$$

We look at the products and the reactants in terms of their entropy and calculate the difference between these two entropy values. ("stoichiometric sum of entropies") (see Fig. 4.6).

Fig. 4.6 Entropy diagram for the autoprotolysis reaction

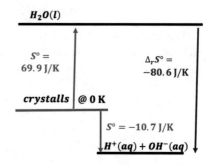

$$\Delta_r S^\circ = (\Delta_f S^\circ (H^+) + \Delta_f S^\circ (OH^-)) - (\Delta_f S^\circ (H_2O)) \qquad (4.37)$$

$$\Delta_r S^\circ = \left(-10.7 \frac{J}{mol\ K}\right) - \left(69.9 \frac{J}{mol\ K}\right) \qquad (4.38)$$

$$\Delta_r S^\circ = -80.6 \frac{J}{mol\ K} \qquad (4.39)$$

During autoprotolysis the entropy decreases strongly (exotropic process).

The second law requires that the entropy of the universe either increases (spontaneous process) or remains constant (reversible process).

Although the second law is valid for the whole universe and our entropy decrease refers only to the system, we may classify exotropic processes as "entropy being not with us."

4.14   How Do We Obtain Free Enthalpy as a Measure of Affinity Using the Laws of Thermodynamics?

First and second law in the general formulation always consider the whole universe: The energy of the universe is constant; the entropy of the universe strives toward a maximum.

If we now want to make statements about the stability of a system, we must concentrate on special cases. For example, for isothermal isobaric processes the combination $(H - TS)$ can only decrease according to GIBBS.

$$\Delta H - T\Delta S \leq 0 \qquad (4.40)$$

We call the combination $H - TS$ Free enthalpy G (or GIBBS' energy)

$$G \equiv H - TS \qquad (4.41)$$

The free enthalpy of a system can only decrease and thus G is a measure of the instability of a system.

$$\Delta G = \Delta H - T\Delta S \qquad (4.42)$$

The change of the free enthalpy ΔG is thus a measure of affinity. Only if ΔG is negative (the so-called exergonic process), an impetus exists; only then the process may run spontaneously.

ΔG has two contributions: the energetic contribution ΔH and the entropic contribution $T\Delta S$.

Free enthalpy of a system can only decrease spontaneously. We can measure the free enthalpy by measuring a reversible work.

Table 4.3 Free enthalpy (GIBBS-energy) factsheet

Free Enthalpy (Gibbs − Energy G)	
G is a measure of	Instability
Statement of the combined First and Second Law	In spontaneous processes, the free enthalpy G of the system can only decrease. At equilibrium, G has a minimum $\Delta G \leq 0$
Measurement of ΔG	Direct measurement as reversible useful work $\Delta G = w_{\text{rev}}$
Calculation of ΔG	From chemical potentials $\Delta G = \mu_{\text{final}} - \mu_{\text{initial}}$ using the Gibbs − Helmholtz − equation $\Delta G = \Delta H - T\Delta S$
G depends on	• Chemical structure • Temperature • Phase • Dilution
Zero point of G	Elements in standard state, 100 kPa, 25 °C
Table values	Free standard enthalpies of formation $\Delta_f G°$ Chemical standard potential $\mu°$
Examples	$\Delta_f G°\,(H_2) = \mu°\,(H_2) = 0\,\frac{kJ}{mol}$ $\Delta_f G°\,(O_2) = \mu°\,(O_2) = 0\,\frac{kJ}{mol}\quad \Delta_f G°\,(H_2O(l)) = \mu°\,(H_2O(l)) = -237.1\,\frac{kJ}{mol}$ $\Delta_f G°\,(H_2O(g)) = \mu°\,(H_2O(g)) = -228.6\,\frac{kJ}{mol}$

$$\Delta G = w_{\text{rev}} \tag{4.43}$$

Much more frequently, however, free enthalpy is calculated applying the so-called GIBBS–HELMHOLTZ equation using ΔH and ΔS (Table 4.3 depicts the free enthalpy factsheet).

$$\Delta G = \Delta H - T\Delta S \tag{4.44}$$

4.15 When Does the Free Enthalpy Change?

Free enthalpy, like entropy, depends on temperature, phase, structure, and dilution, and the zero point for the elements is at 25 °C.

Free enthalpy is a property of an entire system. If free enthalpy is related to just one component of the system, it is also referred to as the "chemical potential" μ of this component.

Using the table values for the standard free enthalpies of various substances [the so-called chemical standard potentials], we can immediately make statements about the instability, for example:

Liquid water has a lower free enthalpy at 25 °C than gaseous water—so it is more stable at 25 °C than gaseous water.

4.16   How Does Free Enthalpy Change During a Reaction?

We can calculate the standard impetus (standard affinity) of the autoprotolysis reaction in different ways. Either we can directly subtract the chemical potentials of reactants and products from each other ("stoichiometric sum of chemical potentials").

$$\Delta_r G^\circ = \mu^\circ_{\text{Products}} - \mu^\circ_{\text{Reactant}} \tag{4.45}$$

$$\Delta_r G^\circ = \left(-157 \frac{\text{kJ}}{\text{mol}}\right) - \left(-237 \frac{\text{kJ}}{\text{mol}}\right) \tag{4.46}$$

$$\Delta_r G^\circ = +80 \frac{\text{kJ}}{\text{mol}} \tag{4.47}$$

Or, since we have already determined the enthalpy of reaction and entropy of reaction, we can plug these two values into the GIBBS–HELMHOLTZ equation:

$$\Delta_r G^\circ = \Delta_r H^\circ - T\Delta_r S^\circ \tag{4.48}$$

$$\Delta_r G^\circ = \left(56 \frac{\text{kJ}}{\text{mol}}\right) - \left(298 \text{ K} \cdot \left(-0.08 \frac{\text{kJ}}{\text{mol K}}\right)\right) \tag{4.49}$$

$$\Delta_r G^\circ = \left(56 \frac{\text{kJ}}{\text{mol}}\right) - \left(-24 \frac{\text{kJ}}{\text{mol}}\right) \tag{4.50}$$

$$\Delta_r G^\circ = 80 \frac{\text{kJ}}{\text{mol}} \tag{4.51}$$

4.17   How Do We Classify a Process Thermodynamically?

We are now able to completely characterize our example process thermodynamically: During autoprotolysis, the energy increases, which means that the process is endothermic—"energy is not with us."

$$H \uparrow: \text{endothermic} \tag{4.52}$$

In autoprotolysis, entropy decreases, that is, the process is exotropic—obviously "entropy is not with us" too.

$$S \downarrow: \text{exotropic} \tag{4.53}$$

If we summarize according to GIBBS–HELMHOLTZ to everything we see that instability increases during the process, which means: the pure products (this is what the standard sign ° in $\mu°$ stands for) are 80 kJ more unstable than the pure reactants.

$$G \uparrow: \text{endergonic} \tag{4.54}$$

The process as a whole is endergonic, the standard impetus (or standard affinity) $\Delta G°$ is positive: water will never spontaneously decompose completely into H^+ and OH^-. However, this does not mean that the process does not occur at all. In the next chapter, we will discuss the difference between the standard affinity $\Delta G°$ and the affinity ΔG and calculate equilibrium constants.

4.18   Summary

The first and the second law of thermodynamics are statements about the energy and the entropy in the universe.

$$\Delta U_{\text{Universe}} = 0 \tag{4.55}$$

$$\Delta S_{\text{Universe}} > 0 \ (\text{spontaneous}) \tag{4.56}$$

We can measure energy change and entropy change during a process

$$\Delta H = q_p \tag{4.57}$$

$$\Delta S = q_{\text{rev}}/T \tag{4.58}$$

or calculate with the help of table values.

$$\Delta_r H° = \Delta_f H° \,(\text{Products}) - \Delta_f H° \,(\text{Reactants}) \tag{4.59}$$

$$\Delta_r S° = S° \,(\text{Products}) - S° \,(\text{Reactants}) \tag{4.60}$$

From the reaction enthalpy and reaction entropy, we can calculate the free enthalpy using the GIBBS–HELMHOLTZ equation

$$\Delta G^\circ = \Delta H^\circ - T\Delta S^\circ \tag{4.61}$$

ΔG° is a measure of affinity, which tells us whether a process can run completely or not.

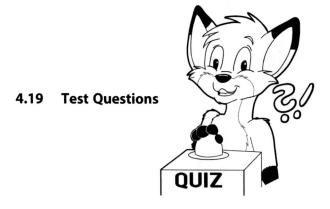

4.19 Test Questions

1. How can we measure enthalpy change ΔH or entropy change ΔS in a process?
   (a) ΔH always corresponds to the isochoric heat of a spontaneous process
   (b) ΔH always corresponds to the isobaric heat of a spontaneous process
   (c) ΔH always corresponds to the reversible heat
   (d) ΔS always corresponds to the reversible heat
   (e) ΔS always corresponds to the reversible reduced heat
   (f) ΔS always corresponds to the isobaric reduced heat
2. Mark the correct statement(s).
   (a) The entropy of a system is either constant or increasing
   (b) The entropy of the universe is either constant or increasing
   (c) The energy of a system is constant or decreasing
   (d) The energy of the universe is constant or decreasing
   (e) Entropy can be generated
   (f) The free enthalpy of an (isobaric, isothermal) system cannot increase spontaneously
   (g) The energy of a system is constant
3. How do enthalpy and entropy change in the following processes? (Do the quantities increase, decrease, or remain the same?)
   (a) 1 mol water is electrolytically decomposed to hydrogen and oxygen
   (b) 1 mol ice melts isobaric to 1 mol water
   (c) 2 moles of hydrogen are mixed isobarically with 1 mole of oxygen to form oxyhydrogen gas
   (d) 1 mol water is isobarically cooled from 25 °C to 0 °C

4.20   Exercises

1. Calculate the heat released when slaking 1 mol of burnt lime.

$$\text{CaO (s)} + \text{H}_2\text{O (l)} \rightarrow \text{Ca(OH)}_2 \text{ (s)} \qquad (4.62)$$

2. Ammonium nitrate can decompose explosively:

$$2\,\text{NH}_4\text{NO}_3 \text{ (s)} \rightarrow 4\text{H}_2\text{O (g)} + 2\,\text{N}_2 \text{ (g)} + \text{O}_2 \text{ (g)} \qquad (4.63)$$

Determine the standard molar enthalpy of reaction $\Delta_r H^\circ$ for the formula conversion formulated here.
Determine the standard molar reaction entropy $\Delta_r S^\circ$ for the formula conversion formulated here.
   Determine the standard free molar enthalpy of reaction $\Delta_r G^\circ$ for the formula conversion formulated here at a temperature of 98.8 °C.
   How much heat (amount) is released during the isobaric decomposition of 1.510 kg ammonium nitrate?

3. Estimate the specific "lower combustion value (net calorific value)" $\Delta_{com} h^\circ$ of ethane (C_2H_6) using bond enthalpies.
   Lower heating value: <u>gaseous</u> water is formed as a combustion product, among other things.

$$2\,\text{C}_2\text{H}_6 \text{ (g)} + 7\,\text{O}_2 \text{ (g)} \rightarrow 4\,\text{CO}_2 \text{ (g)} + 6\,\text{H}_2\text{O (g)} \qquad (4.64)$$

Chemical Equilibrium

5

5.1 Motivation

Many processes do not run to completion, but only up to a certain equilibrium. In this chapter, we learn to calculate this equilibrium and show possibilities to shift the equilibrium. (The *motivational picture* of this chapter, Fig. 5.1, illustrates the effects of energy and entropy change on the position of equilibrium).

5.2 How Do We Quantify the Location of the Equilibrium?

We have seen in the last chapter that the autoprotolysis reaction of water

$$H_2O(l) \rightleftharpoons H^+(aq) + OH^-(aq) \tag{5.1}$$

will never run to completion—this is forbidden by the first and the second law.

In fact, the equilibrium position of the autoprotolysis of water, quantified by the equilibrium constant

$$K_{eq} = \frac{[H^+][OH^-]}{[H_2O]} \tag{5.2}$$

is very far to the left—we will calculate exactly where it is in a moment.

J. S. Lauth, *Physical Chemistry in a Nutshell*, https://doi.org/10.1007/978-3-662-67637-0_5

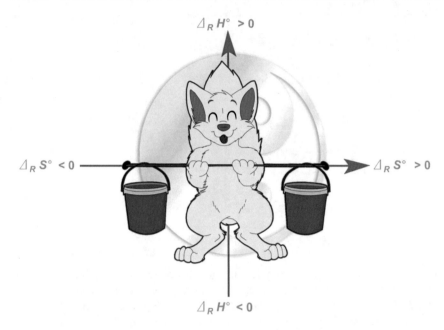

Fig. 5.1 Where does equilibrium lie and how can we shift it? (https://doi.org/10.5446/45979)

$$H_2O(l) \quad \rightleftharpoons \quad H^+(aq) \; + \; OH^-(aq)$$

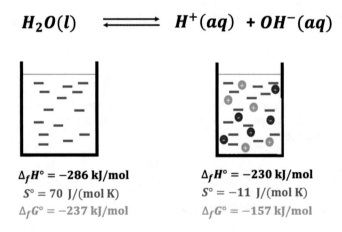

$\Delta_f H° = -286$ kJ/mol
$S° = 70$ J/(mol K)
$\Delta_f G° = -237$ kJ/mol

$\Delta_f H° = -230$ kJ/mol
$S° = -11$ J/(mol K)
$\Delta_f G° = -157$ kJ/mol

Fig. 5.2 Thermodynamic classification of the autoprotolysis reaction

5.3   How Do We Classify a Process with Thermodynamic Parameters?

In the last chapter, we discussed the autoprotolysis of water thermodynamically, i.e. we analyzed the process energetically and entropically (see Fig. 5.2).

The standard enthalpy of reaction $\Delta_r H°$, the standard reaction entropy $\Delta_r S°$, and the free standard reaction enthalpy (or standard affinity or standard impetus) $\Delta_r G°$ always compare the pure products with the pure reactants.

The standard enthalpy of reaction $\Delta_r H°$ of +56 kJ/mol means: the pure products are 56 kJ more energetic than the pure reactants.

The standard reaction entropy $\Delta_r S°$ of −0.08 kJ/K means: the pure products are 0.08 kJ/K lower in entropy than the pure reactants.

Most important thermodynamic parameter is $\Delta_r G°$. The standard impetus $\Delta_r G°$ is +80 kJ; the pure products are 80 kJ more unstable than the pure reactants. This means: the reaction will never run spontaneously to completion.

We now want to see in detail how these three parameters H, S and G change during the reaction.

5.4   Is Energy with Us?

The energy profile of the reaction is relatively simple: The energy [actually: enthalpy] starts at −286 kJ for pure water and ends at −230 kJ for the products. Energy increases rather linearly during the reaction.

On the x-axis, the turnover coefficient ξ is plotted; this indicates the reaction progress: $\xi = 0$ mol means 0% conversion, i.e. exclusively reactants; $\xi = 1$ mol means 100% conversion, i.e. pure products.

Alternatively, we may also indicate the reaction progress using the reaction quotient Q_r which is the quotient of product concentration and reactant concentration.

$$Q_r = \frac{[P]}{[R]} \tag{5.3}$$

Figures 5.3, 5.4, and 5.5 are thermodynamic profiles; they always refer to the reaction of 1 *mol* of reactant. In contrast, in reaction kinetics the reaction profile is related to 1 reactant *molecule*.

Fig. 5.3 Enthalpy profile of the autoprotolysis reaction

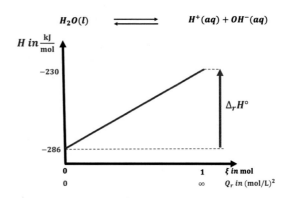

Fig. 5.4 Entropy profile of
the autoprotolysis reaction

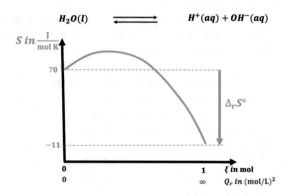

Fig. 5.5 Free enthalpy
profile (chemical potential) of
the autoprotolysis reaction

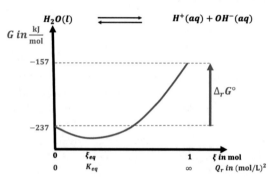

5.5   Is Entropy with Us?

The reaction profile is somewhat different for entropy.

We can first mark the entropy of the pure products and reactants. The entropy
decreases from about 70 J/K for water to −11 J/K for the products.

During reaction a mixture of reactants and products exists; mixtures are usually
more entropic than the pure components. This means that the entropy does not
linearly change from the reactants to the products, but we get a curved curve and
possibly a maximum.

5.6   How Do We Calculate the Standard Impetus (Standard
    Affinity)?

As mentioned in the last chapter, starting with the energetic and entropic numbers of
the reaction and using the GIBBS–HELMHOLTZ equation

$$\Delta_r G^\circ = \Delta_r H^\circ - T \cdot \Delta_r S^\circ \tag{5.4}$$

we are able to calculate the standard impetus $\Delta_r G^\circ$ of the reaction. When using this equation, we must take care to use consistent units (J *or* kJ). When we use $\Delta_r H^\circ = 55.83\,\frac{kJ}{mol}$ we should also use $\Delta_r S^\circ = -0.08061\,\frac{kJ}{K\,mol}$.

$$\Delta_r G^\circ = 55.83\,\frac{kJ}{mol} - 298\ K\left(-0.08061\,\frac{kJ}{K\,mol}\right) \tag{5.5}$$

We get

$$\Delta_r G^\circ = +79.85\,\frac{kJ}{mol} \tag{5.6}$$

5.7   Is Free Enthalpy with Us?

One of the most important thermodynamic figures is the plot of the change in free enthalpy during a process. It tells us how the instability changes during a reaction.

We start with the free standard enthalpy $\Delta_f G^\circ$ (reactant) [other term: chemical standard potential μ° (reactant)] of the reactant at -237 kJ and reach the free standard enthalpy $\Delta_f G^\circ$ (product) [other term: chemical potential μ° (product)] of the products at -157 kJ for complete conversion.

The pure products are therefore more unstable than the pure reactant; the reaction going to completion is thus impossible.

Note that the instability does not increase linearly from reactants to products, but that the curve has a small minimum. From the initial state to this minimum the instability decreases and this means: in this range the reaction can actually take place.

The slope of the curve is a measure of the impetus $\Delta_r G$ (without $^\circ$!)—the more negative the slope $\Delta_r G$, the more impetus the reaction has. In contrast, the standard impetus $\Delta_r G^\circ$ (with $^\circ$!) is the difference of the chemical potentials of products and reactants.

The minimum has a special meaning, because the minimum actually represents the equilibrium. No matter where we start on the curve: the free enthalpy can only decrease, can only approach the minimum.

5.8   How Do We Formulate the Thermodynamic Equilibrium Constant?

You know from general chemistry that an equilibrium

$$R \ \rightleftharpoons \ P \tag{5.7}$$

can be quantified by the law of mass action and by an equilibrium constant

Table 5.1 Thermodynamics convention for specifying concentration $[i]$

Component	Concentration measure	Examples
Gaseous substances	$[i] = p_i$ in bar	100 kPa hydrogen : $[H_2] = 1$ bar 21 kPa oxygen : $[O_2] = 0.21$ bar 2, 3 kPa water vapor : $[H_2O(g)] = 0.023$ bar
Condensed substances (liquids or solids)	$[i] = x_i$ in $\frac{mol}{mol}$	Pure water : $[H_2O(l)] = 1\ \frac{mol}{mol}$ Pure lime : $[CaCO_3(s)] = 1\ \frac{mol}{mol}$ 18 carat gold : $[Au(s)] = 0.6\frac{mol}{mol}$
Dissolved substances	$[i] = c_i$ in $\frac{mol}{L}$	Protons in water : $[H^+(aq)] = 10^{-7}\ \frac{mol}{L}$ Diluted acetic acid : $[HOAc(aq)] = 1\ \frac{mol}{L}$ Dissolved oxygen : $[O_2(aq)] = 0.0003\ \frac{mol}{L}$

$$K_{eq} = \frac{[P]_{eq}}{[R]_{eq}} \tag{5.8}$$

The equilibrium constant looks similar to the reaction quotient

$$Q_r = \frac{[P]}{[R]} \tag{5.9}$$

with the difference that equilibrium concentrations $[P]_{eq}$ and $[R]_{eq}$, respectively, are used. In thermodynamics, it is also important that we follow a convention for the specification of concentration (see Table 5.1):

If we are considering gases, we must use the concentration in bar; if we are considering liquids or solids, we must consider their concentration as mole fraction x, and if we are discussing solutes, we must use the molarity in mol/L.

This means, for our autoprotolysis of water, the equilibrium constant is to be formulated as follows:

$$K_{eq} = \frac{[H^+][OH^-]}{[H_2O]} = \frac{c_{H^+}\, c_{OH^-}}{x_{H_2O}} \tag{5.10}$$

There are two dissolved substances in the numerator of the law of mass action: the concentrations of $[H^+]$ and of $[OH^-]$ are formulated as molarities in mol/L. The denominator contains the concentration of a liquid $[H_2O]$, which must be quantified using the mole fraction.

(Note: The mole fraction of (almost) pure liquids and solids is equal to 1 and can therefore be "omitted.")

For very exact calculations, we have to use the effective concentrations (or activities) of the products and reactants instead of the classically calculated weigh-in concentrations. These are—especially for ions—often calculated with the so-called activity coefficients (keyword: Debye–Hückel theory).

The thermodynamic convention for concentrations determines the unit of the equilibrium constant: $\frac{mol^2}{L^2}$ in our example.

5.9   How Do We Calculate the Thermodynamic Equilibrium Constant?

H^+ and OH^- are about 80 kJ more unstable than water. From this difference in stability (standard impetus $\Delta_r G°$), we can now indeed calculate the numerical value of the equilibrium constant according to this equation.

$$\{K_{eq}\} = \exp\left(-\frac{\Delta_r G°}{RT}\right) \tag{5.11}$$

The curly bracket $\{K_{eq}\}$ means: numerical value of K_{eq}. (More often the square bracket is used around a physical quantity $[K_{eq}]$ means: unit of K_{eq}).

We plug in all quantities and obtain for K_{eq} a numerical value of

$$\{K_{eq}\} = -\exp\left(\frac{79{,}850\,\frac{J}{mol}}{8.314\,\frac{J}{mol\,K}\ 298\ K}\right) = e^{-32} = 1.0 \cdot 10^{-14} \tag{5.12}$$

Together with the unit discussed earlier, we obtain the equilibrium constant of protolysis of water

$$K_{eq} = 1.0 \cdot 10^{-14}\left[\frac{mol}{L}\right]^2 \tag{5.13}$$

Since we are dealing with neutral water

$$[H^+] = [OH^-] \tag{5.14}$$

and the mole fraction of (almost) pure liquids and solids can be set equal to 1, the concentration of protons and thus the pH value can be determined from the equilibrium constant

$$K_{eq} = [H^+][OH^-] \tag{5.15}$$

$$[H^+] = \sqrt{K_{eq}} \tag{5.16}$$

$$pH = -\log\left(\frac{[H^+]}{\frac{mol}{L}}\right) \tag{5.17}$$

At 25 °C, this results in a pH value of 7 for neutral water.

$$[H^+] = \sqrt{1.0 \cdot 10^{-14} \left[\frac{mol}{L}\right]^2} = 1.0 \cdot 10^{-7} \frac{mol}{L} \qquad (5.18)$$

$$pH = -\log\left(\frac{1.0 \cdot 10^{-7} \frac{mol}{L}}{\frac{mol}{L}}\right) = 7.0 \qquad (5.19)$$

5.10   How Do We Classify a Process in an Entropy/Enthalpy Diagram?

In general, we can classify any process thermodynamically with respect to the change in energy and the change in entropy. We now want to plot the entropy change $\Delta_{rxn}S$ as the abscissa and the enthalpy change $\Delta_{rxn}H$ as the ordinate in a coordinate system. Then we can discuss the four combinations in the four quadrants and formulate example reactions (see Fig. 5.6).

Processes in which energy and entropy are "with us," i.e. energy decreases ($\Delta_{rxn}H < 0$) and the entropy increases ($\Delta_{rxn}S > 0$), are always exergonic, always have an impetus, e.g. the endotropic and exothermic decomposition of ammonium nitrate

$$2\,NH_4NO_3 \rightleftharpoons 4\,H_2O\,(g) + 2\,N_2\,(g) + O_2(g) \qquad (5.20)$$

On the other hand, processes with both energy and entropy being "not with us" ($\Delta_{rxn}S < 0$; $\Delta_{rxn}H > 0$) are always endergonic. The exotropic endothermic transformation of graphite into diamond is an example here.

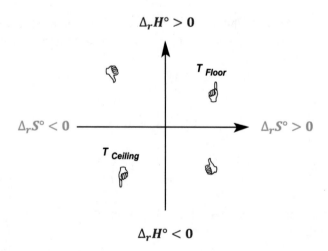

Fig. 5.6 $\Delta H°$ and $\Delta S°$ combinations

$$C(s, \text{graphite}) \rightleftharpoons C(s, \text{diamond}) \tag{5.21}$$

In many reactions impetus switches with change in temperature.
The decomposition of N_2O_4 into NO_2

$$N_2O_4(g) \rightleftharpoons 2\,NO_2\,(g) \tag{5.22}$$

is endothermic ($\Delta_{rxn}H < 0$; "energy is not with us"), but endotropic ($\Delta_{rxn}S > 0$; "entropy is with us"). This means that at low temperature, where energy has the upper hand, the reaction does not have a standard impetus; but at high temperature, it does. The threshold temperature is called the "floor temperature."

$$T_{\text{floor}} = \frac{\Delta_{rxn}H}{\Delta_{rxn}S} \tag{5.23}$$

Conversely, for reactions in the lower left quadrant, "energy is with us" ($\Delta_{rxn}H < 0$) but "entropy is not with us" ($\Delta_{rxn}S < 0$). Since temperature fights on the side of entropy, this means that at low temperature there is standard impetus (so equilibrium is on the right) and at high temperature there is no standard impetus (equilibrium is on the left). The threshold temperature is a "ceiling temperature."

$$T_{\text{ceiling}} = \frac{\Delta_{rxn}H}{\Delta_{rxn}S} \tag{5.24}$$

5.11   How Does Temperature Change Standard Impetus and Equilibrium Constant?

As we have seen, temperature has a great influence both on standard impetus $\Delta_{rxn}G°$ and on the equilibrium constant $\{K_{eq}\}$. We now want to quantify this influence. We consider the GIBBS–HELMHOLTZ equation, which quantifies the influence of temperature on standard impetus

$$\Delta_r G° = \Delta_r H° - T\Delta_r S \tag{5.25}$$

and we consider the equation with which we can calculate the equilibrium constant—temperature also is involved in here.

$$\ln\{K_{eq}\} = -\frac{\Delta_r G°}{RT} \tag{5.26}$$

If we combine these two equations, we get the following expression, which describes the function $\{K_{eq}\} = f(T)$.

$$\ln\{K_{eq}\} = -\frac{\Delta_r H°}{R}\frac{1}{T} + \frac{\Delta_r S°}{R} \tag{5.27}$$

Fig. 5.7 VAN'T HOFF reaction isobars for endothermic and exothermic reactions

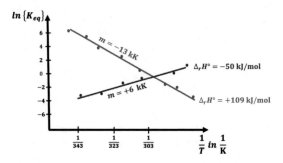

We see that the equilibrium constant does not depend on temperature in a simple way. We can establish a linear relationship from this equation if we plot ln $(\{K_{eq}\})$ against the reciprocal of the absolute temperature $1/T$—this is the so-called VAN'T HOFF reaction isobar.

The VAN'T HOFF reaction isobar is a straight line. In Fig. 5.7, the equilibrium constants of two different reactions have been measured at different temperatures and plotted according to VAN'T HOFF.

Note that the equilibrium constant increases with temperature in one case (blue curve) and decreases with temperature in the other case (red curve). According to VAN'T HOFF'S equation, the enthalpy of reaction must be negative in the first case (exothermic reaction); in the second case, we are dealing with an endothermic reaction.

We can also use VAN'T HOFF's equation to convert equilibrium constants from one temperature to another.

$$\ln\left(\frac{K'_{eq}}{K_{eq}}\right) = -\frac{\Delta_r H}{R}\left(\frac{1}{T'} - \frac{1}{T}\right) \tag{5.28}$$

5.12 How Can We Change the Position of an Equilibrium?

Qualitatively, we may state from Fig. 5.7 that the equilibrium constant of an endothermic reaction increases with increasing temperature.

The reverse is true for exothermic reactions.

Extending this statement, LE CHATELIER and BRAUN formulated the general principle of least constraint for equilibria:

When we apply a constraint to a system in equilibrium, that equilibrium will shift by avoiding the constraint—by consuming an offered quantity.

We want to use an example—the decay of N_2O_4

$$N_2O_4 \,(g) \rightleftharpoons 2\,NO_2\,(g) \tag{5.29}$$

to explain this principle.

The position of this special equilibrium can be easily detected optically. N_2O_4 is a colorless gas, NO_2 is a brown colored gas.

The reaction from N_2O_4 (g) to NO_2 (g) is endothermic, thus consumes heat. If we increase temperature—thus offering heat—the equilibrium will shift in the direction in which the heat is consumed, i.e. to the right-hand side. Quantitatively we have already discussed this using VAN'T HOFF'S equation: endothermic reactions shift their equilibrium at higher temperature to the right-hand side.

$$\Delta_r H > 0: \quad T \uparrow \Rightarrow \{K_{eq}\} \uparrow \tag{5.30}$$

$$\Delta_r H < 0: \quad T \uparrow \Rightarrow \{K_{eq}\} \downarrow \tag{5.31}$$

At the process N_2O_4 (g) to NO_2 (g) the volume increases—it is an endochoric process. If we now increase the pressure, the equilibrium shifts in the direction where volume is smaller, i.e. it shifts to the left.

$$\Delta_r V > 0: \quad p \uparrow \quad \Rightarrow \text{product yield} \downarrow \tag{5.32}$$

$$\Delta_r V < 0: \quad p \uparrow \quad \Rightarrow \text{product yield} \uparrow \tag{5.33}$$

5.13   How Can We Provoke Endergonic Reactions?

There are also reactions that are endergonic at any temperature, so they never have a standard impetus. What can we do in this case?

We may illustrate this reaction with a mechanical analog: A weight lying on the ground will never voluntarily move upward—at any temperature! This process of lifting upwards has no impetus. If we want to provoke this process, we have to supply energy from outside. The energy must not be provided as heat, but as useful work, e.g. by connecting an electric motor in the mechanical case. Such a stubbornly endergonic reaction is, e.g., the conversion of CO_2 and water into glucose and oxygen.

$$6\,CO_2(g) + 6\,H_2O\,(g) \rightarrow C_6H_{12}O_6(s) + 6\,O_2\,(g) \tag{5.34}$$

This so-called photosynthesis only works because light energy—i.e., no heat—is fed into the system from outside and can be coupled into the reaction process.

Likewise, table salt will not voluntarily split into sodium and chlorine.

$$NaCl(s) \rightarrow Na(l) + \frac{1}{2}\,Cl_2\,(g) \tag{5.35}$$

This is only possible if we apply electrical energy from outside by means of electrolysis.

There is another way to lift the weight in our mechanical analog. We have to couple another process, which has a large impetus, to the stubbornly endergonic process. For example, in our mechanical analog, this can be done by moving another weight down from a certain height—this is voluntary, this is spontaneous—and coupling this exergonic process to our first weight via a pulley. Coupling multiple reactions is very common in biochemistry. The ATP $\rightarrow$ ADP reaction is a high impetus exergonic reaction that likes to be coupled in.

We may also provoke the endergonic decomposition of iron oxide to iron

$$Fe_2O_3 \rightarrow 2 \, Fe \, (l) + \frac{3}{2} \, O_2 \, (g) \tag{5.36}$$

by removing the oxygen out of equilibrium using another reaction, e.g. with aluminum or magnesium.

5.14 Summary

We have seen that the impetus of a process has two portions—an energetic portion $\Delta_{Rxn}H°$ and an entropic portion $\Delta_{Rxn}S°$.

$$\Delta_r G° = \Delta_r H° - T \cdot \Delta_r S° \tag{5.37}$$

The effect of $\Delta_r S°$ will always be enhanced by temperature. Magnitude and sign of $\Delta_r H°$ and $\Delta_r S°$ and possibly temperature control the impetus of a process.

The standard impetus $\Delta_r G°$ describes the difference of instability between reactants and products. Using $\Delta_r G°$ we can calculate the equilibrium constant K_{eq}

$$\ln\{K_{eq}\} = -\frac{\Delta_r G°}{RT} \tag{5.38}$$

The temperature dependence of the equilibrium constants is described by the equation of VAN 'T HOFF.

$$\ln\{K_{eq}\} = -\frac{\Delta_r H°}{R} \frac{1}{T} + \frac{\Delta_r S°}{R} \tag{5.39}$$

Qualitatively, we can discuss shifting of equilibria with the principle of least constraint.

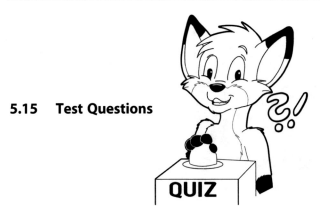

5.15 Test Questions

1. Mark the correct statement(s)
   (a) Exothermic endotropic processes are always exergonic
   (b) Exothermic exotropic processes are exergonic below the ceiling temperature
   (c) Endothermic endotropic processes are endergonic above the floor temperature

2. Which measures regarding pressure and temperature favor the yield of product in the following reversible reactions?
   (a) CO_2 (g)+ C (s) $\rightleftharpoons$ 2 CO (g) (endothermic)
   (b) H_2 (g)+$H_2C{=}CH_2$ (g) $\rightleftharpoons$ $H_3C{-}CH_3$ (g) (exothermic)
   (c) CO_2 (g) $\rightleftharpoons$ CO_2 (aq) (exothermic and exochoric)

3. The neutralization reaction is exothermic and endotropic. The pH of pure water at 25 °C is 7.0.
   What is the pH of pure water at 37 °C?
   (a) pH > 7.0
   (b) pH = 7.0
   (c) pH < 7.0

4. Why are spontaneous processes sometimes called "downhill" in free enthalpy?
   (a) Equilibrium is reached whenever both enthalpy and entropy decrease.
   (b) The difference in free enthalpy between a product and a reactant is always negative.
   (c) An equilibrium corresponds to a state of minimum free energy.

5.16   Exercises

1. Discuss the "water gas shift reaction."

$$CO(g) + H_2O\ (g) \rightleftharpoons CO_2\ (g) + H_2\ (g) \qquad (5.40)$$

(a) Is the reaction to carbon dioxide at 25 °C endothermic or exothermic?
(b) Is the reaction to carbon dioxide at 25 °C endergonic or exergonic?

2. At what temperature T_{floor} does calcium carbonate ($CaCO_3$) start to decompose ($p = 100$ kPa)? For the calculation, use the data for 25 °C (the so-called ULICH approximation)?

$$CaCO_3(s) \rightleftharpoons CaO\ (s) + CO_2\ (g) \qquad (5.41)$$

3. Calculate the standard affinity of the neutralization reaction $\Delta_r G°$ at 25 °C.

$$H^+(aq) + OH^-\ (aq) \rightleftharpoons H_2O\ (l) \qquad (5.42)$$

4. Calculate the equilibrium constant K_{GG} for the BOUDOUARD equilibrium

$$CO_2(g) + C\ (s) \rightleftharpoons 2\ CO\ (g) \qquad (5.43)$$

at 500 °C

Vapor Pressure

6

6.1 Motivation

Liquids can evaporate; gases can condense. At what pressures and temperatures do they do so?

Quantitative description of phase transitions is essential for understanding and designing separation methods such as distillation, extraction, or absorption. (The *motivational picture* of this chapter, Fig. 6.1, illustrates the phase diagram of carbon dioxide).

6.2 What Is Vapor Pressure?

In this chapter, we start the topic "Phase equilibria" and will mainly discuss vapor pressure.

What is vapor pressure anyway?

Consider an evacuated vessel at 20 °C and we put liquid water into it. Initially, the pressure in the vessel (or system as the thermodynamicist says) will be 0 kPa (see Fig. 6.2 top). However, the water will begin to evaporate and the pressure above the liquid phase will increase. Partial pressure of water vapor builds up in the gas phase—depicted in the second picture

$$p_{H_2O} = 1.15 \text{ kPa} \qquad (6.1)$$

Water will continue to evaporate and at 2.34 kPa the pressure above the liquid phase will remain constant. We are now at equilibrium—at phase equilibrium. We

J. S. Lauth, *Physical Chemistry in a Nutshell*,
https://doi.org/10.1007/978-3-662-67637-0_6

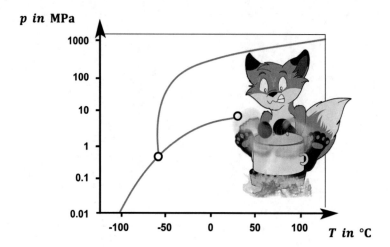

Fig. 6.1 How do we describe phase equilibria? (https://doi.org/10.5446/46038)

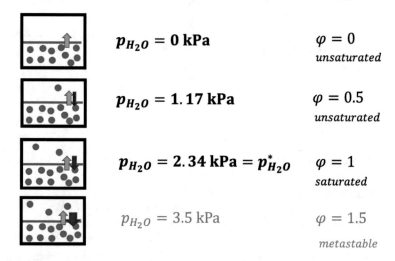

Fig. 6.2 Origin of vapor pressure of water at 20 °C

call the equilibrium partial pressure above a condensed phase vapor pressure p^*. The vapor pressure of water at 20 °C is

$$p_{H_2O}(20\,°C) = 2.34\,\text{kPa} = p^*_{H_2O} \tag{6.2}$$

The index $*$ is to remind us that we are dealing with vapor pressure.

Vapor pressure is strongly dependent on temperature. Liquid water at 0 °C shows a vapor pressure p^*

$$p^*_{H_2O}(0\,^{\circ}C) = 0.612 \text{ kPa} \tag{6.3}$$

And at 100 °C the vapor pressure of water will be

$$p^*_{H_2O}(100\,^{\circ}C) = 101.3 \text{ kPa} \tag{6.4}$$

101.3 kPa is exactly 1.00 atm and thus about standard external pressure. When vapor pressure of a liquid equals external pressure, the liquid will start to boil; 100 °C is the normal boiling temperature of water.

If we relate the partial pressure of water to the vapor pressure of water, we obtain the relative humidity φ.

$$\varphi = \frac{p_{H_2O}}{p^*_{H_2O}} \tag{6.5}$$

This relative humidity is 1.0 or 100% when vapor pressure is reached. In the second picture, for example, relative humidity is 50%, in the first picture relative humidity is 0%. The gas phase can even be super-saturated in water vapor, then the humidity $\varphi > 1$ (metastable state).

6.3 When Are Two Phases in Equilibrium?

Vapor pressure is a special case for phase equilibrium.

$$\alpha \rightarrow \beta$$
$$H_2O \text{ (Phase } \alpha) \quad \rightleftharpoons \quad H_2O \text{ (Phase } \beta) \tag{6.6}$$
$$\beta \rightarrow \alpha$$

In every phase equilibrium there is at least one component which can transfer the phase boundary, i.e. can change between two phases. We call this component "transition component" (see Fig. 6.3).

Fig. 6.3 Vapor pressure of water as dynamic equilibrium of condensation and evaporation

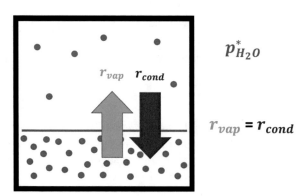

$$p^*_{H_2O}$$

$$r_{vap} = r_{cond}$$

What is the condition for equilibrium in this case? We can consider the situation thermodynamically: Equilibrium prevails whenever the water molecules feel equally comfortable in both phases, when the chemical potential—the instability—of water is the same in both phases.

$$\mu^\alpha_{H_2O} = \mu^\beta_{H_2O} \tag{6.7}$$

This is the general thermodynamic condition for phase equilibria: the chemical potential of the transition component in one phase is equal to the chemical potential of the transition component in the other phase.

We can also consider the question kinetically: Equilibrium exists when the rate of transition of the component from phase α to phase β—in our example, the evaporation rate of water—is equal to the rate of transition of the component from phase β back into phase α—in our example, the condensation rate.

$$r(\alpha \rightarrow \beta) = r(\beta \rightarrow \alpha) \tag{6.8}$$

This is an example of a so-called dynamic equilibrium. Macroscopically, we see no change, but microscopically, a forward reaction and a backward reaction occur at the same rate.

6.4   What Factors Do Affect Vapor Pressure?

Vapor pressure of a substance depends primarily on the volatility of this substance, quantified by its enthalpy of vaporization

$$p^*_i = f\left(\Delta_{vap}H^\circ\right) \tag{6.9}$$

Water therefore has a lower vapor pressure than a low boiler such as ethanol.

$$p^*_{H_2O}(20\,^\circ C) = 2.34 \text{ kPa} \tag{6.10}$$

$$p^*_{C_2H_5OH}(20\,^\circ C) = 5.8 \text{ kPa} \tag{6.11}$$

In fact, we can take vapor pressure as a measure of the "escape tendency" of a molecule from the liquid phase.

As already briefly mentioned, vapor pressure depends primarily on the temperature.

$$p^*_i = f(T) \tag{6.12}$$

As a rule of thumb, vapor pressure almost doubles for every 10 °C (10 K, 18 °F) increase in temperature. This corresponds to an exponential function, as we will see in a moment (CLAUSIUS–CLAPEYRON equation).

Vapor pressure also depends on the purity of the phase. If we mix a second substance to our water, vapor pressure will drop.

$$p_i^* = f(x) \tag{6.13}$$

The rule of thumb here is: 1% impurity content lowers vapor pressure by 1% (*RAOULT's 1st* law).

The vapor pressure also depends to a small extent on whether we have a flat surface or a curved surface. For small droplets, for example, vapor pressure is a little bit larger than for a flat surface (*KELVIN* equation).

$$p_i^* = f(r) \tag{6.14}$$

Vapor pressure also increases when an inert gas is additionally present at high pressure.

$$p_i^* = f(p_{\text{inert}}) \tag{6.15}$$

6.5   What Does the Vapor Pressure Diagram of a Pure Substance Look Like?

We remember the three-dimensional $p\overline{V}\,T$ phase diagram. Of course, we can visualize vapor pressure here (see Fig. 6.4).

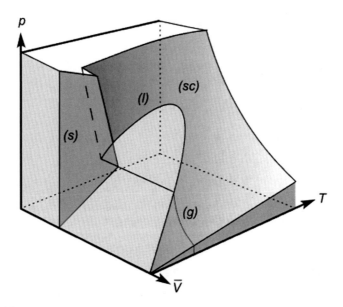

Fig. 6.4 $p\overline{V}T$ phase diagram of water [(H$_2$O), one-component system]. s solid, l liquid, g gaseous, sc supercritical

Fig. 6.5 pT diagram of water with triple point (T) and critical point (C) as well as standard melting point (F) and standard boiling point (V)

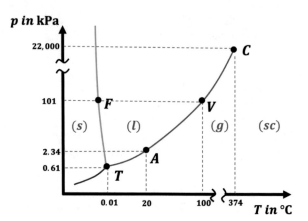

It is the realm, in which liquid and gas coexist, bordered from the homogeneous realms by a binodal in the form of an inverted parabola. If we project the $p\overline{V}\,T$ phase diagram onto the pT surface, this two-phase realm is condensed to a line and we obtain a very common two-dimensional vapor pressure diagram of a pure substance, which we will discuss below.

We identify three curves in this diagram, which are the vapor pressure curve, the sublimation pressure curve, and the melting pressure curve. These three lines meet in one point—the so-called triple point. The triple point for water is at 0.01 °C and at 0.612 kPa and is a fixed point given by nature, which is suitable, e.g., for calibrating thermometers.

$$p^*_{H_2O}(0.01\,°C) = 0.612 \text{ kPa} \tag{6.16}$$

The triple point marks the beginning of the vapor pressure curve. The vapor pressure curve rises to the right and ends at the critical point. The critical point of water is at 374 °C and 22,000 kPa.

$$p^*_{H_2O}(374\,°C) = 22 \text{ MPa} \tag{6.17}$$

We can read boiling points of the liquid for any pressures in this diagram. For example, the "normal" boiling point of water at 101 kPa is 100 °C.

$$p^*_{H_2O}(100\,°C) = 101.3 \text{ kPa} \tag{6.18}$$

At 20 °C, vapor pressure of water is 2.34 kPa; if we had a pressure of 2.34 kPa as external pressure, water would already boil at 20 °C.

$$p^*_{H_2O}(20\,°C) = 2.34 \text{ kPa} \tag{6.19}$$

By converse we can also use Fig. 6.5 to determine the temperature at which gaseous water condenses (called the dew point). At the dew point, vapor pressure is equal to partial pressure.

6.6   How Can We Describe the Vapor Pressure Curve Mathematically?

There has been no lack of attempts to describe the vapor curve mathematically. In the equation named after them, CLAUSIUSandCLAPEYRON related two points on the vapor pressure curve $(T, p^{*\,\prime}\,;\,T',\,p^{*\prime})$ via the enthalpy of vaporization $\Delta_{vap}H°$.

$$\ln\left(\frac{p^{*\prime}}{p^*}\right) = -\frac{\Delta_{vap}H°}{R}\left(\frac{1}{T'} - \frac{1}{T}\right) \qquad (6.20)$$

Even more popular for calculating vapor pressures is ANTOINE's equation which assigns three ANTOINE factors A, B, and C to each liquid (see appendix for detailed table).

$$\log\left(\frac{p^*}{kPa}\right) = A - \frac{B}{C+T} \qquad (6.21)$$

It gives very precise results for vapor pressures, but does not have as nice a theoretical background as CLAUSIUS–CLAPEYRON's equation.

6.7   How Can We Evaluate the Vapor Pressure Curve?

With the help of CLAUSIUS–CLAPEYRON's equation, we can also evaluate a vapor pressure curve: Plotting the logarithm of the vapor pressure against the reciprocal of the absolute temperature will give a straight line with a negative slope (see Fig. 6.6).

From the slope, we can determine the enthalpy of vaporization $\Delta_{vap}H°$ (or heat of vaporization). The enthalpy of vaporization of water at 25 °C is about 40 kJ/mol (for

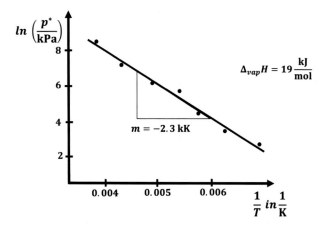

Fig. 6.6 CLAUSIUS–CLAPEYRON plot of the vapor pressure curve and determination of the molar enthalpy of vaporization

comparison: average thermal energy at 25 °C ~4 kJ/mol; bond enthalpy of the OH bond: ~400 kJ/mol).

6.8   Do Solids Show Vapor Pressure?

When we cool liquid water from 20 °C to 0 °C, the vapor pressure drops to about a quarter of its original value.

$$p^*_{H_2O,l}(0\,^\circ C) = 0.61 \text{ kPa} \tag{6.22}$$

We can also cool liquid water below 0 °C, e.g. to -10 °C; then we measure a vapor pressure of only

$$p^*_{H_2O,l}(-10\,^\circ C) = 0.29 \text{ kPa} \tag{6.23}$$

This value lies on the red curve (left-hand side) in Fig. 6.7.
Solid ice at -10 °C has a lower vapor pressure—viz.

$$p^*_{ice}(-10\,^\circ C) = 0.26 \text{ kPa} \tag{6.24}$$

This value is on the blue curve in Fig. 6.7 (on the right-hand side). Low vapor pressure means that water molecules are more comfortable in this phase, they have a lower chemical potential, they are less unstable.

At 0 °C, ice and liquid water show the same vapor pressure—at 0 °C, both phases are equally stable.

$$p^*_{ice}(0\,^\circ C) = 0.61 \text{ kPa} \tag{6.25}$$

Fig. 6.7 Vapor pressure curves of liquid and solid water with triple point (section of the pT diagram)

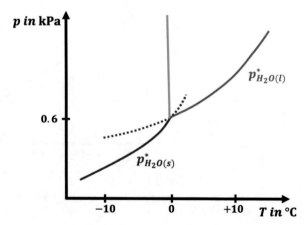

6.9   How Can We Describe the Composition of a Mixture?

What happens when we dissolve a substance in our solvent water? (see Fig. 6.8)

Before we deal with the properties of the mixture, we first need to clearly describe the composition of the mixture. For example, we take 1.00 kg of water as solvent (abbreviated as A) and 1.00 mol of sugar as solute (abbreviated as B).

$$m_A = 1.00 \text{ kg} \tag{6.26}$$

$$n_B = 1.00 \text{ mol} \tag{6.27}$$

We obtain a homogeneous mixture A/B, which contains about four times as much sugar as apple juice or iced tea. Now we can calculate, for example, the mole fraction x of this mixture by dividing the amount of solute B by the total amount of solute.

$$x_B = \frac{n_B}{n_B + n_A} \tag{6.28}$$

Our solution does have a molar fraction of 1.80 mol% sugar.

$$x_{sugar} = 1.8 \text{ mol\%} \tag{6.29}$$

Indicating concentration using molarity c is very common in chemistry. To calculate this quantity, we divide the amount of solute by the total volume.

$$c_B = \frac{n_B}{V_{total}} \tag{6.30}$$

$$c_{sugar} = 0.82 \frac{\text{mol}}{\text{L}} \tag{6.31}$$

Molarity c should not be confused with molality b: to calculate b we have to divide the amount of solute by the mass of the solvent A.

Fig. 6.8 Preparation of a homogeneous mixture of components A and B

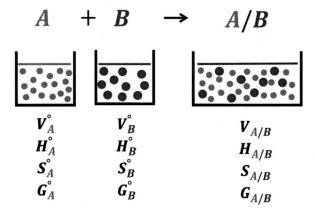

$$b_B = \frac{n_B}{m_A} \tag{6.32}$$

$$b_{sugar} = 1.00 \frac{mol}{kg} \tag{6.33}$$

In very dilute solutions, molarity and molality are approximately equal. We obtain such a dilute solution, for example, if we dissolve 10 g of carbon dioxide in 1 kg of water (this corresponds to classic sparkling water). We obtain a solution with molarity 0.23 mol/L and molality 0.23 mol/kg.

For solutes that dissociate, the actual number of dissolved particles is relevant, quantified by VAN'T HOFF'S factor i. This is then referred to as osmolarity $i \cdot c$, osmolality $i \cdot b$,and osmolar fraction $i \cdot x$.

6.10   How Is the Solvent Distributed Between the Liquid and Gas Phases?

Now, how does vapor pressure of a solution compare to vapor pressure of a solvent? Let us take this iced tea, for example, i.e. a solution of sugar in water, and again consider the phase equilibrium between the liquid and gaseous phases (see Fig. 6.9).

Water molecules can change between liquid phase and gas phase.

$$H_2O^l \rightleftharpoons H_2O^g \tag{6.34}$$

As for any equilibrium, the law of mass action can be formulated

$$\frac{[H_2O^g]}{[H_2O^l]} = K_{eq} \tag{6.35}$$

We have to express the concentration of a gas phase as pressure (in bar); the concentration of a liquid as its mole fraction.

We end up with RAOULT'S 1ST law, which states in words: The concentration of the solvent molecules in the gas phase divided by the concentration of the solvent molecules in the liquid phase is a constant.

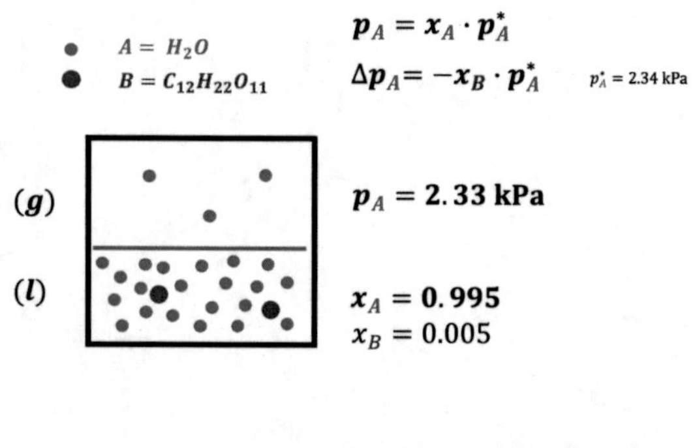

Fig. 6.9 Phase equilibrium of the solvent between liquid phase and gas phase using the example of iced tea to illustrate RAOULT's 1st law

• $A = H_2O$
• $B = C_{12}H_{22}O_{11}$

(g)

(l)

$p_A = x_A \cdot p_A^*$

$\Delta p_A = -x_B \cdot p_A^*$ $p_A^* = 2.34\ kPa$

$p_A = 2.33\ kPa$

$x_A = 0.995$
$x_B = 0.005$

$$\frac{p_A}{x_A} = p_A^* \tag{6.36}$$

With pure water $x_A = 1$ and we obtain the classical vapor pressure of water

$$p_{H_2O}^*(20\,^\circ C) = 2.34 \text{ kPa} \tag{6.37}$$

If, as in the iced tea example, there is a 0.5% sugar solution, the liquid phase is only 99.5% water and the vapor pressure above this solution is also reduced by 0.5% to

$$p_{H_2O. \text{ sugar solution}}^*(20\,^\circ C) = 2.27 \text{ kPa} \tag{6.38}$$

RAOULT's law applies strictly only to ideal solutions, i.e. solutions in which A and B are chemically similar and thus indifferent to each other.

If components A and B are energetically attracted to or repelled from each other, negative or positive deviations from RAOULT's law will occur.

6.11   How Is the Solute Distributed Between the Liquid and Gas Phases?

If we do not dissolve the low-volatile sugar but a gas in water, e.g. carbon dioxide in a lemonade, then another phase equilibrium will occur, because the dissolved substance B—in our case CO_2—can also transfer back and forth between two phases (see Fig. 6.10).

We can describe the equilibrium for the dissolved component in the same way as the equilibrium for the solvent.

$$CO_2{}^l \rightleftharpoons CO_2{}^g \tag{6.39}$$

We may formulate the mass action law.

Fig. 6.10 Phase equilibrium of the solute between the liquid phase and the gas phase using the example of lemonade to illustrate HENRY's Law

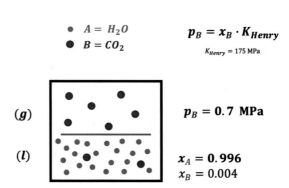

$\bullet \; A = H_2O$

$\bullet \; B = CO_2$

$p_B = x_B \cdot K_{Henry}$

$K_{Henry} = 175$ MPa

(g)

(l)

$p_B = 0.7$ MPa

$x_A = 0.996$

$x_B = 0.004$

$$\frac{[CO_2{}^g]}{[CO_2{}^l]} = K_{eq} \tag{6.40}$$

In words: The pressure of CO_2 in the gas phase divided by the mole fraction of CO_2 in the liquid phase is a constant.

$$\frac{p_B}{x_B} = K_{Henry} \tag{6.41}$$

This equation is *HENRY's* law of absorption and this states just like *RAOULT's* 1st law that the phases are equally occupied by components. If the concentration in phase α increases, it must also increase in phase β ("partition law").

We can use *HENRY's* law to calculate the pressure in a lemonade bottle. We need the mole fraction of CO_2 in the liquid phase (which is 0.4 mol%) and we need *HENRY's* constant for CO_2 in water at 20 °C

$$K_{Henry}(CO_2) = 175 \text{ MPa} \tag{6.42}$$

We calculate a pressure of 0.7 MPa or 7 bar in the gas phase.

6.12   How Does a Solute Partition Between Two Liquid Phases?

For the sake of completeness, we can also mention another phase equilibrium here, namely the distributive equilibrium described by *NERNST*.

Consider two liquids that do not mix, e.g. water and oil. We consider a third component that can transfer between these two solvents as a transition component, e.g. acetic acid between water and oil (see Fig. 6.11).

We formulate the equilibrium as in *RAOULT's* or *HENRY's* law

Fig. 6.11 Phase equilibrium of the transition component between two solvents (raffinate and extractant) to illustrate NERNST's distribution law

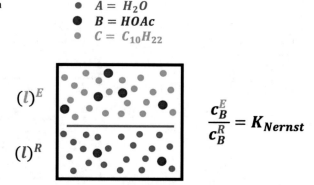

$$B^\alpha \rightleftharpoons B^\beta \tag{6.43}$$

And the corresponding law of mass action

$$\frac{[B]^\beta}{[B]^\alpha} = K_{eq} \tag{6.44}$$

The concentration of acetic acid in the oil phase divided by the concentration of acetic acid in the aqueous phase is a constant; the phases have parity with the transition component.

$$\frac{c_B^E}{c_B^R} = K_{Nernst} \tag{6.45}$$

NERNST's distribution law is the basic law for understanding the separation method extraction.

6.13   Summary

Vapor pressure is the equilibrium partial pressure above a condensed phase. Vapor pressure primarily depends on the volatility of the substance and on temperature.

$$p_{H_2O}^* = f(\Delta_{vap}H^\circ, T, x) \tag{6.46}$$

The vapor pressure curve of a pure substance, which starts at the triple point and ends at the critical point, can be described, for example, according to *CLAUSIUS–CLAPEYRON*.

$$\ln\left(\frac{p^{*'}}{p^*}\right) = -\frac{\Delta_{vap}H^\circ}{R}\left(\frac{1}{T'} - \frac{1}{T}\right) \tag{6.47}$$

Vapor pressure of a solution will be lower than the vapor pressure of the pure solvent—mathematically described by *RAOULT's 1ST* law.

$$\frac{p_A}{x_A} = p_A^*$$ (6.48)

When we dissolve a gas in a liquid, the amount of gas dissolved is proportional to the partial pressure of the gas above the liquid—mathematically described by HENRY's law of absorption.

$$\frac{p_B}{x_B} = K_{Henry}$$ (6.49)

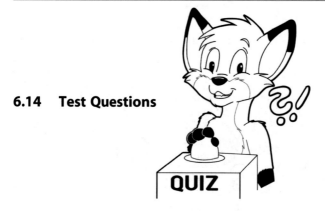

6.14   Test Questions

1. How does vapor pressure react when
   (a) ... temperature rises?
   (b) ...volume of the gas phase increases?
   (c) ...the surface has a negative curvature (drop)?
   (d) ...another liquid is added (ideal mixture)?
2. Mark the correct statement(s).
   (a) Supercooled liquid water has a lower vapor pressure at $-1\,°C$ than ice at $-1\,°C$
   (b) Pure water has a lower vapor pressure than a salt solution
   (c) A small drop of water has a lower vapor pressure than a large drop of water
   (d) Iodine has a lower vapor pressure in a vacuum than in the presence of 100 bar nitrogen (inert gas)
   (e) The higher the HENRY constant, the better the gas solubility

3. Figure 6.12 shows the phase diagram of carbon dioxide.
   Mark the correct statement(s).

Fig. 6.12 Phase diagram (*pT* diagram) of carbon dioxide

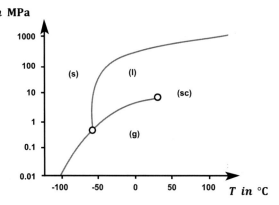

(a) Carbon dioxide can be liquefied by pressure at 0 °C.
(b) At standard pressure (100 kPa) there can only be solid or gaseous carbon dioxide
(c) Carbon dioxide can be liquefied by pressure at 50 °C.
(d) The melting point of solid carbon dioxide at triple pressure is higher than at 10 MPa

6.15 Exercises

Use ANTOINE'S or CLAUSIUS–CLAPEYRON'S equation for the calculations.

1. How much oxygen can dissolve in 1.00 kg of water at 25 °C [in equilibrium with air ($p_{O2} = 21$ kPa)]?
   (K_{HENRY} (O$_2$ in water, 25 °C) = 4.6 GPa)
2. At what temperature does acetone boil at an external pressure of 72 kPa?
3. Determine the boiling temperature of ethanol at 50 kPa.
4. The vapor pressure of a liquid is measured at different temperatures

Temperature	Vapor pressure
3.73 °C	0.743 kPa
10.59 °C	1.30 kPa

(a) Calculate the molar enthalpy of vaporization of the liquid.
(b) What would be the boiling point of the liquid at standard pressure (100 kPa).

5. A flue gas has a temperature of 100.0 °C and a relative humidity of 20.0% at 100 kPa.
   Calculate the dew point of the gas mixture.

Solutions

7

7.1 Motivation

When we dissolve a substance in a solvent, its properties change. We will discuss important properties of solutions quantitatively in this chapter. (The *motivational picture* of this chapter, Fig. 7.1, illustrates the effect of osmotic pressure of a solution).

7.2 How Can We Specify the Composition of a Mixture?

Before we start with the actual topic, a note on the concentration of solutions, especially salt solutions (see Fig. 7.2 and Table 7.1).

For example, if we dissolve 1.00 mole of sugar in 1.00 kg of water, we have a molar solution: There are 1 mole of dissolved particles in 1 kg of water. This number of dissolved particles is what matters for the colligative properties.

$$b_B = \frac{n_B}{m_A} \tag{7.1}$$

$$b_{sugar} = 1.0 \frac{mol}{kg} \tag{7.2}$$

J. S. Lauth, *Physical Chemistry in a Nutshell*,
https://doi.org/10.1007/978-3-662-67637-0_7

Fig. 7.1 How do solvent and solution differ? (https://doi.org/10.5446/46039)

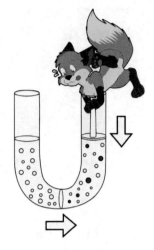

Fig. 7.2 Preparation of a homogeneous mixture of components A and B

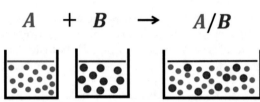

Table 7.1 Important measures of concentration

(Os-)	Mole fraction	$x_B = \frac{n_B}{n_A + n_B}$	$(\cdot i)$
(Os-)	Molarity	$c_B = \frac{n_B}{V_{total}}$	$(\cdot i)$
(Os-)	Molality	$b_B = \frac{n_B}{m_A}$	$(\cdot i)$
	Mass concentration	$\gamma_B = \frac{m_B}{V_{total}}$	

7.3   How Can We Specify the Composition of an Electrolyte Solution?

Therefore, when we make a salt solution, we must consider if and how the salt dissociates.

$$K_{\nu^+}A_{\nu^-} \xrightarrow{\alpha} \nu^+ K^{z^+} + \nu^- A^{z^-} \tag{7.3}$$

ν^+ and ν^- are the decay numbers, z^+ and z^- are the charge numbers of the ions, α is the degree of dissociation.

For example, half a mole of common salt produces 1 mole of particles in solution—half a mole of cations and half a mole of anions.

$$NaCl \rightarrow Na^+ + Cl^- \tag{7.4}$$

The number of dissolved particles is twice the amount of NaCl dissolved in the first place. This ratio of the number of dissolved particles to the number of dissolved moles of salt is quantified by the VAN'T HOFF factor i.

$$i = (\nu_+ + \nu_- - 1)\,\alpha + 1 \tag{7.5}$$

For completely dissociated sodium chloride.

$$i = 2 \tag{7.6}$$

Do not confuse the VAN'T HOFF *factor I with the electrochemical* coefficient n_e. *The latter indicates how many positive charges are in 1 mol of electrolyte.*

$$n_e = \nu^+ z^+ = |\nu^- z^-| \tag{7.7}$$

Osmolarity $i \cdot b_B$ and osmolality $i \cdot c_B$ were introduced to quantify the number of particles in the concentration specification.

$$i \cdot b_B = i \cdot \frac{n_B}{m_A} \tag{7.8}$$

$$i \cdot c_B = i \cdot \frac{n_B}{V_{total}} \tag{7.9}$$

Our solution has a molality of

$$b_{NaCl} = 0.50\,\frac{mol}{kg} \tag{7.10}$$

but an osmolality of

$$i \cdot b_{NaCl} = 1.00\,\frac{mol}{kg} \tag{7.11}$$

Thus, the salt solution contains the same amount of dissolved particles as the sugar solution discussed previously. The two solutions have equal osmolality; the two solutions are isotonic.

Analogously, we obtain the osmolarity from the molarity by multiplying by the VAN'T HOFF-factor i.

7.4   How Well Do Two Components A and B Get Along?

A simple solution consists of two components: the solvent A and the solute B. Depending on how the interactions between these components are, we distinguish between ideal and real solutions.

If we mix two chemically very similar substances, the enthalpy of the mixture is the same as the enthalpy of the initial substances; during mixing neither heat is released nor absorbed; the enthalpy of mixing $\Delta_{mix}H°$ is zero.

$$\text{MeOH(l)} + \text{EtOH(l)} \rightarrow \text{MeOH/EtOH(l)} \tag{7.12}$$

$$\Delta_{mix}H° = 0\,\frac{\text{kJ}}{\text{mol}} \tag{7.13}$$

We speak here of ideal solutions, the so-called FLORY–HUGGINS coefficient, which quantifies the energetic interactions between the different components, is zero.

$$\chi = 0 \tag{7.14}$$

However, if the two components are energetically "attracting" or "repelling," heat is released or absorbed during mixing. Real solutions may possess negative or positive FLORY–HUGGINS coefficients.

$$\text{HCl(conc.)} + \text{H}_2\text{O(l)} \rightarrow \text{HCl (dil.)} \tag{7.15}$$

$$\Delta_{dil}H° = -10\,\frac{\text{kJ}}{\text{mol}} \tag{7.16}$$

$$\chi < 0 \tag{7.17}$$

$$\text{KCl(s)} + \text{H}_2\text{O(l)} \rightarrow \text{KCl(aq)} \tag{7.18}$$

$$\Delta_{solv}H° = +17\,\frac{\text{kJ}}{\text{mol}} \tag{7.19}$$

$$\chi > 0 \tag{7.20}$$

7.5   How Do We Describe a Mixing Process Thermodynamically?

FLORY and HUGGINS analyzed the mixture of two components thermodynamically. Depending on the magnitude of the intra- and inter-component interactions, the mixture will be either ideal, exothermic, or endothermic, expressed by the FLORY–HUGGINS-coefficient χ. The theory derives equations to calculate both enthalpy of mixing and entropy of mixing.

$$\Delta_{mix}H = RT\,\chi\,x_A\,x_B \tag{7.21}$$

$$\Delta_{\text{mix}}S = -R\left(x_A \ln\left(x_A\right) + x_B \ln\left(x_B\right)\right) \tag{7.22}$$

For detailed discussions, the free enthalpy of mixing is of particular interest. The theory is able to predict the conditions for good or bad miscibility and for miscibility gaps.

$$\Delta_{\text{mix}}G = RT \chi\, x_A\, x_B + RT\left(x_A \ln\left(x_A\right) + x_B \ln\left(x_B\right)\right) \tag{7.23}$$

The following statements apply to ideal solutions, i.e., solutions in which intra- and inter-component interactions are similar—χ is then equal to 0; miscibility gaps do not exist here.

7.6   At What Temperature Does a Solution Boil?

The upper curves in Fig. 7.3 show the vapor pressure curve, melting pressure curve, and triple point of the pure solvent. The corresponding vapor pressure curve for the solution is plotted below. According to RAOULT'S 1st law, vapor pressure of a solution is always lower than the vapor pressure of the solvent.

$$\frac{p_A}{x_A} = p_A^* \tag{7.24}$$

Reformulating and taking into account VAN'T HOFF factor, we obtain for the vapor pressure lowering

$$\Delta p_A = -x_B \cdot p_A^* \cdot i \tag{7.25}$$

Fig. 7.3 Vapor pressure curve of solvent and solution to illustrate lowering of vapor pressure and elevation of boiling point

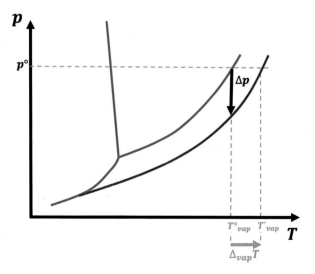

A solvent boils when its vapor pressure is equal to the external pressure. However, a solution has a lower vapor pressure at the same temperature—as just discussed. To make the solution boil as well, we need to increase temperature: A solution therefore starts to boil at a higher temperature compared to the solvent; quantified by the boiling point elevation.

$$\Delta_{vap}T = k_{eb} \cdot b_B \cdot i \tag{7.26}$$

k_{eb} is the ebullioscopic constant, a characteristic for each solvent. For water this constant is

$$k_{eb\ (H_2O)} = 0.512 \frac{K\ kg}{mol} \tag{7.27}$$

This means our 1.0 molar sugar solution and our 1.0 osmolar salt solution both have the identical boiling point of 100.5 °C.

7.7   At What Temperature Does a Solution Freeze? (see Fig. 7.4)

A solvent freezes when liquid and solid have the same vapor pressure, i.e. at the temperature where the vapor pressure curve and the sublimation pressure curve intersect (triple point). With a solution, this intersection is shifted to lower temperatures—there is a freezing point depression of a solution compared to the solvent, quantified by RAOULT'S 2ND law.

$$\Delta_{fus}T = -k_{kr} \cdot b_B \cdot i \tag{7.28}$$

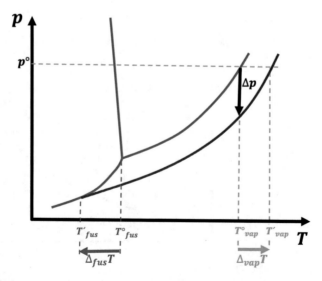

Fig. 7.4 Vapor pressure curve of solvent and solution to illustrate vapor pressure lowering and freezing point depression

The constant k_{kr} is the cryoscopic constant, another characteristic for each solvent. For water the cryoscopic constant is

$$k_{kr\ (H_2O)} = 1.86\ \frac{K\ kg}{mol} \tag{7.29}$$

Both our sugar solution and salt solution freeze at $-1.86\ °C$.

Solutions with the same osmolarity, i.e. identical number of dissolved particles, are called "isotonic." Isotonic solutions agree in all colligative properties.

We have already learned about three of the four colligative properties: vapor pressure lowering, boiling point elevation, freezing point depression. The fourth colligative property, osmotic pressure, is particularly important.

7.8   Why Does the Solvent Migrate into the More Concentrated Solution?

A solution always has a lower vapor pressure than the solvent, which means the solution is more stable than the solvent (see Fig. 7.5).

If we separate a solution and a solvent through a semipermeable membrane (permeable only to the solvent), then solvent molecules will voluntarily migrate into the more concentrated solution. This causes an overpressure to build up in the solution.

This transport process continues until the pressure in the solution has become so high that the stability of both phases is identical again. A so-called osmotic pressure will build up in the more concentrated solution; we can calculate it according to VAN'T HOFF using osmolarity, gas constant, and temperature.

$$\Pi = c_B \cdot R \cdot T \cdot i \tag{7.30}$$

Fig. 7.5 Origin of osmotic pressure at the semipermeable phase boundary between a concentrated and dilute solution

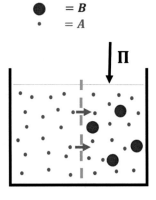

7.9   Where Do We Encounter Osmotic Pressure in Nature and Technology?

A commercially available [physiological] saline solution, for example, has an osmolarity of 0.3 mol/kg; its osmotic pressure is 0.76 MPa [7.6 bar]—just like the osmotic pressure of human blood. Since many cell membranes are semipermeable, osmotic pressure plays a major role in biology. Body fluids should be isotonic, otherwise cells could be damaged (see Fig. 7.6).

In hypertonic media, for example, the cells can lose water very quickly and thus die. This also means that in concentrated solutions no microorganisms can persist—this is the basis of preservation by pickling or candying and therefore the dead sea has its name (see Fig. 7.7).

It is also possible to reverse osmosis: By applying an appropriately high pressure, a solution can be ultrafiltrated; in this way, seawater can be desalinated to produce fresh water.

Osmotic pressure, freezing point depression, and boiling point elevation are colligative properties whose magnitude depend only on the number of dissolved particles, not on their nature. Therefore, these properties are well suited for

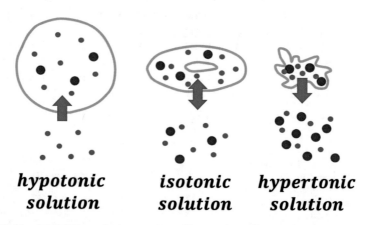

hypotonic   *isotonic*   *hypertonic*
solution   *solution*   *solution*

Fig. 7.6 Blood cells in isotonic, hypertonic, and hypotonic media

Fig. 7.7 Principle of reverse osmosis [e.g., used for desalination of seawater (right-hand side)]

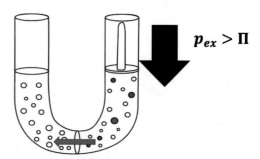

$p_{ex} > \Pi$

determining molar masses of solutes. Osmometrically, the molar masses of proteins or polymers can be determined.

7.10 Summary

Thermodynamics of mixing can be described according to *FLORY–HUGGINS*; especially for ideal solutions there is no heat of mixing and no volume of mixing. The *FLORY–HUGGINS parameter* $\chi = 0$.

$$\Delta_{mix}H = RT \, \chi \, x_A \, x_B \tag{7.31}$$

$$\Delta_{mix}S = -R \left(x_A \, \ln \left(x_A \right) + x_B \, \ln \left(x_B \right) \right) \tag{7.32}$$

A solution always has a lower vapor pressure than a pure solvent. The vapor pressure lowering can be calculated according to *RAOULT'S* 1st law.

$$\Delta p_A = -x_B \cdot p_A^* \cdot i \tag{7.33}$$

A solution has a higher boiling point and a lower freezing point than the pure solvent.

$$\Delta_{fus}T = -k_{kr} \cdot b_B \cdot i \tag{7.34}$$

$$\Delta_{vap}T = k_{eb} \cdot b_B \cdot i \tag{7.35}$$

The mentioned three properties are colligative, which means: only the number of dissolved particles is relevant, not the type of dissolved particles. Osmotic pressure is another colligative property; it can be calculated according to *VAN'T HOFF* and it plays a major role especially in biology.

$$\Pi = c_B \cdot R \cdot T \cdot i \tag{7.36}$$

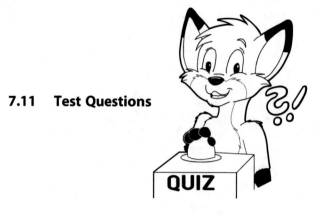

7.11   Test Questions

1. Which solution has the lowest boiling point?
   Which solution has the lowest freezing point?
   (a) 0.050 mol/kg $CaCl_2$ solution ($i = 3$)
   (b) 0.15 mol/kg NaCl solution ($i = 2$)
   (c) 0.10 mol/kg HCl solution (hydrochloric acid, $i = 2$)
   (d) 0.050 mol/kg CH_3COOH (acetic acid, $i = 1.1$)
   (e) 0.20 mol/kg $C_{12}H_{22}O_{11}$ (sucrose, $i = 1$)

2. Mark the correct statement(s)
   Compared to the solvent, a solution has...
   (a) ...a higher melting point
   (b) ...a higher boiling point
   (c) ...a higher vapor pressure

3. Components A and B form an ideal mixture

Initial state Final state
   (component A + component B) → (mixture of components)
   Mark the correct statement(s)
   In the preparation of an ideal mixture of the components...
   (a) the enthalpy of the system increases
   (b) the entropy of the system increases
   (c) the free enthalpy of the system increases
   (d) the volume of the system increases

7.12   Exercises

1. The aqueous solution of a protein (3.50 mg protein in 5.00 mL solution) shows an osmotic pressure of 205 Pa at 25.0 °C.
   Determine the molar mass of the protein ($i = 1$).
2. The average osmotic pressure of blood is 780 kPa at 37.0 °C.
   What is the molarity of a glucose solution ($C_6H_{12}O_6$) that is isotonic with blood?
3. 60.90 g urea (NH_2-CO-NH_2, $M = 60.06$ g/mol) are dissolved in 0.500 kg water.
   The density of the solution is 1.000 kg/L
   Urea does not dissociate in water and forms an ideal solution with water. Pure water freezes at 0.00 °C and has a vapor pressure of 101.325 kPa at 100 °C.
   The cryoscopic constant of water is 1.86 K kg/mol.
   (a) Determine the freezing point of the solution?
   (b) Find the osmotic pressure of the solution at 11.2 °C?
   (c) What is the vapor pressure of the solution at 100 °C?
4. 11.23 g of common salt (NaCl, $M = 58.44$ g/mol, $i = 2$) are dissolved in 1.00 kg of water. The volume of the solution is 1.00 L.
   (a) What is the molarity of the solution?
   (b) What is the osmolarity of the solution?
   (c) What is the osmotic pressure of the solution at 34.3 °C?

Phase Diagrams

8

8.1   Motivation

The behavior of pure substances or mixtures is very often described graphically using phase diagrams. How can we translate these diagrams into real life? (The *motivational picture* of this chapter, Fig. 8.1, illustrates the phase diagram of an ideal mixture of two components).

8.2   How Do We Describe a Two-Component System?

This chapter is about the interpretation of phase diagrams, in particular the meaning of the binodals, tie lines, and invariant points occurring in them (see Fig. 8.2).

Isopropyl alcohol (IPA) and isobutyl alcohol (IBA) are chemically very similar, so both components form ideal mixtures. This two-component system will guide us through the chapter as an example.

We mix 1 mole of IPA and 1 mole of IBA and heat this mixture to 92 °C. The total pressure will be 100 kPa then. If we look closely at the gas phase, we find that it is 70% IPA, while the liquid phase is only 50% IPA. This is ONE possible state of the two-component IBA/IPA system, characterized by temperature, pressure, and composition.

© The Author(s), under exclusive license to Springer-Verlag GmbH, DE, part of
Springer Nature 2023
J. S. Lauth, *Physical Chemistry in a Nutshell*,
https://doi.org/10.1007/978-3-662-67637-0_8

Fig. 8.1 How do we read phase diagrams of multicomponent systems? (https://doi.org/10.5446/ 45975)

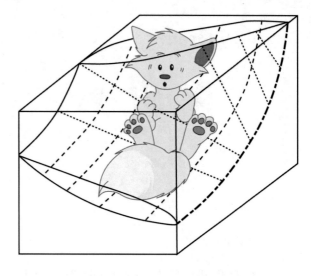

Fig. 8.2 Phase equilibrium liquid/gas for an IPA-IBA mixture

$$T = 92\,°C \quad p = 100\,kPa$$

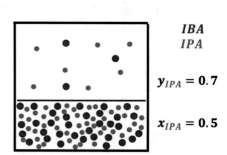

IBA
IPA

$$y_{IPA} = 0.7$$

$$x_{IPA} = 0.5$$

8.3   What Does the Phase Diagram of a Two-Component System (2CS) Look like?

If we want to plot all possible states of this 2-C system, we need a three-dimensional diagram with temperature, pressure, and composition as axes. We may get this kind of representation (see Fig. 8.3).

A typical feature of the phase diagram of a 2-C system is the confined x-axis: the concentration of B can only take values from 0 to 100%.

Note: Both mass % and mole % are commonly used in the literature to indicate concentration. These two values are clearly different for some systems.

The three state variables of the system (temperature, pressure and composition) correspond to a *point* in the phase diagram. Depending on where this *point* is located, we can identify which phases are present. When discussing a phase diagram, it is useful to first mark the areas that correspond to a single phase (homogeneous realm),

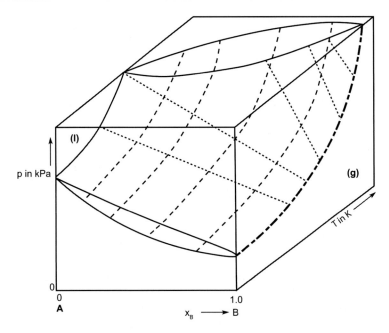

Fig. 8.3 pTx phase diagram of a mixture (plane in front: vapor pressure diagram; plane on top: boiling diagram)

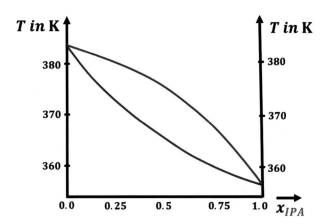

Fig. 8.4 Boiling point diagram of the system IBA(2-methylpropanol-1)/IPA(propanol-2)

e.g. the liquid phase realm and the gas phase realm. The remaining regions are then multiphase.

For example, inside the "tube" in Fig. 8.4 lies the liquid/gaseous region (l/g). The separation lines or surfaces between homogeneous and heterogeneous regions are called binodals (also called "coexistence curves," "binodal lines," "binodal curves," "binodal surfaces," respectively).

In Fig. 8.4, two binodal surfaces exist: The two-phase region is separated upward from the liquid-phase realm by the boiling point surface, and downward from the gaseous realm by the dew point surface.

8.4   Where Do We Find a 2D Boiling Point Diagram in the 3D Phase Diagram?

For the sake of clarity, intersections (projections) of the phase diagram in Fig. 8.4 are very often chosen—either at constant temperature or at constant pressure. These are the cross sectional surfaces at the top or front of the 3D diagram in Fig. 8.4.

The binodal surfaces then become binodal lines on these 2D diagrams: boiling point curve and dew point curve are homogeneously falling or rising in ideal mixtures.

In the following, we want to discuss exclusively the boiling point diagram of the system IBA/IPA, i.e. the top surface of the 3D diagram. We select standard pressure $p°$ as the constant pressure and end up with a phase diagram in which composition acts as the x-axis and temperature as the y-axis.

On the x-axis, we will find the pure high-boiling component ("heavy ends," IBA) on the left-hand side and the pure low-boiling component ("light ends," IPA) on the right-hand side. The two binodals boiling point curve and dew point curve meet in two points—these are the so-called invariant points. At invariant points, a phase change takes place at constant temperature.

8.5   At What Temperature Does a Liquid Mixture Start to Boil?

We can now ask ourselves at what temperature a 50:50 mixture IBA/IPA will start to boil (see Fig. 8.5).

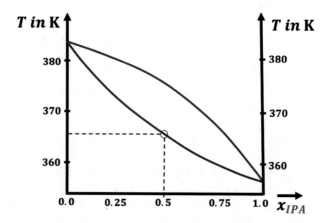

Fig. 8.5 Discussion of the boiling point diagram of the system IBA/IPA (I): Determination of the boiling temperature of a liquid mixture

To do this, we draw a so-called isopleth—a line on which the composition of the mixture is constant, in the diagram and look where this line meets the boiling point curve. The point of intersection is at 365 K—this is obviously where our 50:50 mixture will start to boil.

8.6 At What Temperature Does a Gaseous Mixture Start to Condense?

Our phase diagram also tells us at what temperature a 50:50 gaseous mixture of IBA and IPA will start to condense (see Fig. 8.6).

We only have to draw the isopleth starting from high temperatures and mark the intersection with the dew point curve. Obviously the intersection is located at 371 K: At 371 K, the first liquid droplets will condense from a gaseous 50:50 mixture.

8.7 What Is the Composition of the Gas Phase Above a Boiling Mixture?

Let us reconsider the liquid 50:50 mixture. This mixture boils at 365 K. What does the gas phase above this boiling mixture look like? Well, to answer this question, we have to "consult the tie line."

A tie line is a straight line in the heterogeneous region of the diagram that connects the two phases that are in equilibrium with each other. As all equilibrium lines, tie lines are always both isotherms and isobars, and there are in principle an infinite number of them (see Fig. 8.7).

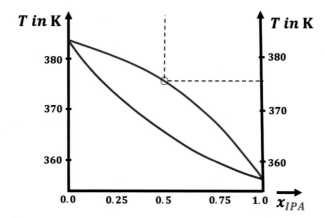

Fig. 8.6 Discussion of the boiling point diagram of the system IBA/IPA (II): Determination of the condensation temperature of a gaseous mixture

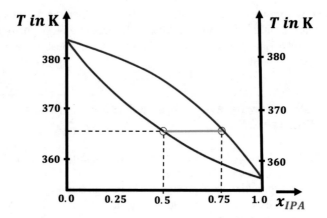

Fig. 8.7 Discussion of the boiling point diagram of the system IBA/IPA (III): Determination of the composition of gas and liquid phase during boiling (tie line)

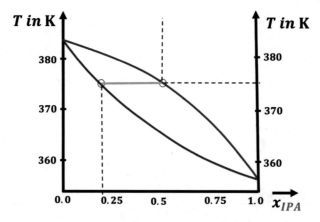

Fig. 8.8 Discussion of the boiling point diagram of the system IBA/IPA (IV): Determination of the composition of gas and liquid phase during condensation (tie line)

We draw a tie line starting at the isopleth/boiling point curve intersection across the two-phase region until we reach the dew point curve. The intersection with the dew point curve is at $x_{IPA} = 0.7$, which is the composition of the gas phase. The gas phase above a 50:50 liquid mixture is thus enriched with the low-boiling component.

8.8   What Is the Composition of the Liquid Phase Condensing from a Gas Phase?

Similarly, we can use the phase diagram to determine that the liquid formed from a 50:50 gas phase mixture will be enriched with the high-boiling component. The appropriately drawn tie line tells us that the first liquid droplets contain about 20% IPA (see Fig. 8.8).

8.9   Heterogeneous Regions in Phase Diagrams: Which Phases Are Present and in What Quantities?

Using the phase diagram we can discuss arbitrary temperatures and compositions (see Fig. 8.9).

For example, if we heat a 50:50 mixture to 370 K, the system will be located in the two-phase region, which means: the mixture we have chosen is unstable in homogeneous form, but will "decompose" along the tie line into a liquid phase and a gas phase.

The intersections of the tie line with the binodals tell us the compositions of these phases. The liquid phase will be 40% IPA and the gas phase will be 55% IPA. If we want to determine the amounts of liquid and gas phase, we need to apply the so-called lever rule.

The lever arm a multiplied by the amount of liquid phase is equal to the lever arm b multiplied by the amount of gas phase. In our example, a is about twice as large as b, which means: there will be twice as much gas phase as liquid phase.

$$a \cdot n_{\text{liquid}} = b \cdot n_{\text{gas}} \tag{8.1}$$

8.10   How Do We Read the Boiling Point Diagram of a Non-ideal Mixture?

The example of isobutyl alcohol/isopropyl alcohol just discussed corresponds to an ideal mixture. In non-ideal ("real") mixtures, maxima or minima can occur in the phase diagram, an example being the well-known system ethanol/water (see Fig. 8.10).

Boiling point curve and dew point curve now meet three times—not only in the pure components, but also in a minimum. This minimum is called azeotrope. So there are three invariant points in this diagram: the boiling point of pure water at a constant 100 °C (212 °F, 373 K), the boiling point of pure ethanol at a constant 78.3 °C (172.9 °F, 351.5 K), and the boiling point of the azeotrope at a constant 78.2 °C (172.8 °F, 351.4 K). An azeotrope boils and condenses like a pure substance and cannot be separated into its components by distillation.

Fig. 8.9 Discussion of the boiling point diagram of the system IBA/IPA (V): Determination of the quantity ratio of gas and liquid phase (lever rule)

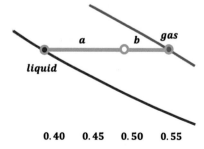

| | 0.40 | 0.45 | 0.50 | 0.55 |

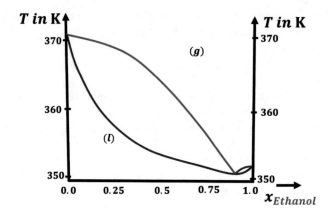

Fig. 8.10 Minimum azeotrope for the system ethanol/water

$$H_2O(l) \quad \overset{100\,^\circ C}{\rightleftharpoons} \quad H_2O(l) \tag{8.2}$$

$$C_2H_5OH(l) \quad \overset{78.3\,^\circ C}{\rightleftharpoons} \quad C_2H_5OH(l) \tag{8.3}$$

$$\text{azeotrope}(95.6\%, l) \quad \overset{78.2\,^\circ C}{\rightleftharpoons} \quad \text{azeotrope}(95.6\%, g) \tag{8.4}$$

8.11   How Do We Read the Melting Point Diagram of a Solid Solution System?

Figure 8.11 is also a phase diagram of an ideal 2-C system; however, it is not a boiling point diagram but a melting point diagram.

The binodals here are called the solidus line and the liquidus line, and they tell us when a solid phase melts or a liquid phase solidifies.

As with any ideal phase diagram, there are only two invariant points, namely the phase transitions of the pure substances, in this case the melting point of silver and the melting point of gold.

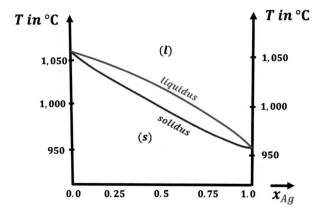

Fig. 8.11 Melting point diagram of an ideal system (silver/gold)

$$962\,^{\circ}\mathrm{C}$$
$$\mathrm{Ag\,(s)} \quad \rightleftharpoons \quad \mathrm{Ag\,(l)} \tag{8.5}$$

$$1064\,^{\circ}\mathrm{C}$$
$$\mathrm{Au\,(s)} \quad \rightleftharpoons \quad \mathrm{Au(l)} \tag{8.6}$$

8.12 What Does the Melting Point Diagram Look Like When the Solid Phase Has a Miscibility Gap?

Figure 8.12 shows the phase diagram, or more precisely: the melting point diagram, of a non-ideal mixture, the system silver/copper. There are three invariant points, besides the melting points of silver and copper

$$962\,^{\circ}\mathrm{C}$$
$$\mathrm{Ag\,(s)} \quad \rightleftharpoons \quad \mathrm{Ag\,(l)} \tag{8.7}$$

$$1085\,^{\circ}\mathrm{C}$$
$$\mathrm{Cu\,(s)} \quad \rightleftharpoons \quad \mathrm{Cu(l)} \tag{8.8}$$

We recognize a characteristic V-shaped minimum in the liquidus line. It is called a eutectic and is the lowest melting mixture of this system. A eutectic melts and

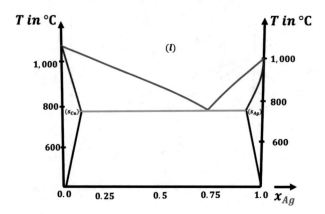

Fig. 8.12 Melting point diagram with eutectic (silver/copper)

solidifies like a pure substance at a constant temperature. There are three phases involved in the eutectic, two solids and the melt. The melting of a eutectic can be formulated as follows:

$$\overset{T_{\text{eutectic}}}{(s)^{I} + (s)^{II} \quad \rightleftarrows \quad (l)} \tag{8.9}$$

A homogeneous liquid phase is formed from two solid phases I and II; the temperature is constant during the phase transition.

$$\overset{780\,^{\circ}C}{\text{Ag (s)} + \text{Cu (s)} \quad \rightleftharpoons \quad \text{eutectic} \,(72\%, l)} \tag{8.10}$$

8.13   What Is Incongruent Melting?

Figure 8.13 shows the phase diagram of the system water/table salt. It is somewhat more complicated than the diagrams shown so far because, in addition to ice and solid table salt, there is another homogeneous solid phase, halite. Halite, chemical formula $NaCl \cdot 2H_2O$, is stable only up to about 0 °C and then decomposes into solid NaCl and brine.

We consistently choose our standard approach when discussing this phase diagram:

First, we mark the *homogeneous realms* in the diagram:

These are liquid phase (brine), solid H_2O (ice), solid halite, and solid NaCl. All other regions are heterogeneous.

Then we name the *binodals* and important *tie lines*:

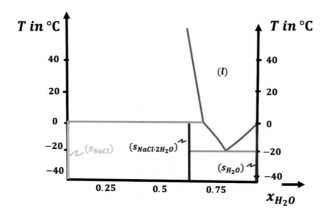

Fig. 8.13   Melting diagram with peritectic and eutectic (water/NaCl)

The tie line describing the decay of halite and the tie line passing through the eutectic, the so-called eutectic.

Finally, we look for the *invariant points* and specify the phase transitions that will take place at these points:

The decomposition of halite at 0 °C corresponds in a certain way to the inversion of a eutectic; this is called a peritectic or incongruent melting.

A peritectic also involves three phases, a solid phase I breaks down into a liquid phase and another solid phase II.

$$(s)^I \quad \underset{}{\overset{T_{\text{peritectic}}}{\rightleftarrows}} \quad (l) + (s)^{II} \tag{8.11}$$

The temperature is constant during the phase transition.

$$0\,°C$$
$$\text{NaCl} \cdot 2\text{H}_2\text{O (halite, s)} \;\rightleftharpoons\; \text{brine } (70\%, l) + \text{NaCl (s)} \tag{8.12}$$

8.14   How Do We Represent Three-Component Systems Graphically?

If we mix three components, we can specify any mixtures graphically in an equilateral triangle. This idea dates back to GIBBS—the so-called GIBBS phase triangle (Fig. 8.14).

Fig. 8.14 GIBBS phase triangle

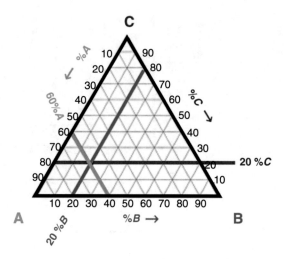

Fig. 8.15 GIBBS phase triangle of chloroform/water/ acetic acid with miscibility gap

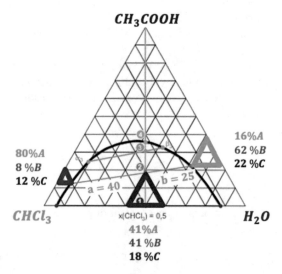

The corners of the triangle represent the pure components, the edges of the triangle represent two-component mixtures, and the inside of the triangle corresponds to three-component mixtures.

When discussing the phase triangle, it is important to note that the composition axes here are not perpendicular to each other, but take 60° angles.

The composition of the dot marked in Fig. 8.15, for example, is 20% component C, 20% component B, and the remainder, i.e. 60% component A.

8.15 How Do Binodals and Tie Lines Run in G$_{IBBS}$ Phase Triangle?

The phase triangle may include both homogeneous and heterogeneous regions and thus there may also be binodals and tie lines.

Figure 9.1 shows the phase triangle of the three solvents chloroform, water, and acetic acid. Below the binodal, the system is heterogeneous, above the system is homogeneous.

If we start with a 50:50 mixture of chloroform/water, the system will be located in the heterogeneous region at point (1). If we then gradually add more and more acetic acid, we arrive at points (2), (3), and (4). We can draw tie lines through points (2) and (3) which tell us what phases are actually present. In point (2), for example, there is an organic phase of composition α_2 (little auxiliary triangle) and a largely aqueous phase of composition β_2 (midsized auxiliary triangle).

We can give the exact composition of the phases α_2 and β_2—here an auxiliary triangle construction has proved useful—and we can also give the quantity ratio of the phases α_2 and β_2 by applying the lever law. Lever arm a is larger than lever arm b; this means that the β_2 is present in excess.

8.16 Summary

Phase diagrams are best discussed by first finding the homogeneous and heterogeneous regions, then labeling and naming the binodals, and drawing in some tie lines. After that, we can still name the invariant points and specify the processes that occur at these points.

Ideal phase diagrams, whether melting point or boiling point diagrams, show binodals with neither maximum nor minimum.

Non-ideal ("real") mixtures may show maxima and minima, which are called azeotrope, peritectic, or eutectic, depending on the type of diagram.

For the illustration of 3-component mixtures, G$_{IBBS}$ phase triangle diagram has proved useful. In this diagram, too, there can be binodals and tie lines.

A tie line tells us not only the composition of the phases in the heterogeneous region, but also their quantitative ratio; we have to apply the lever rule for this purpose.

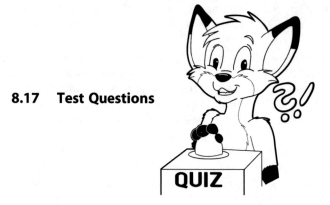

8.17   Test Questions

1. Mark the correct statement(s)
   When boiling an ideal two-component mixture,...
   (a) the boiling temperature remains constant
   (b) the composition of the gas phase remains constant
   (c) the gas phase is always enriched with low boilers
   (d) the composition of the liquid phase remains constant

2. During solidification of a eutectic melt....
   (a) the temperature remains constant
   (b) the composition of the melt remains constant
   (c) two fixed phases are created
   (d) the cooling curve shows a breakpoint
   (e) a homogeneous solid solution is formed as the solid phase

3. Which invariant points do you recognize in the phase diagram of the two-component system water/table salt (see Fig. 8.16).

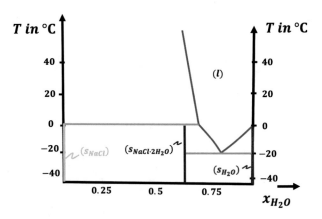

Fig. 8.16 Phase diagram of the system H₂O/NaCl

8.18   Exercises

1. Consider the phase diagram of the system LiCl/KCl (see Fig. 8.17)

(a) A solid mixture containing 40 mol% LiCl is heated—starting at 20 °C—and begins to melt. What is the temperature and what is the composition of the melt?

(b) A liquid mixture with 40% LiCl content is cooled—starting at 800 °C—and begins to solidify. What is the temperature and what is the composition of the solid phase?

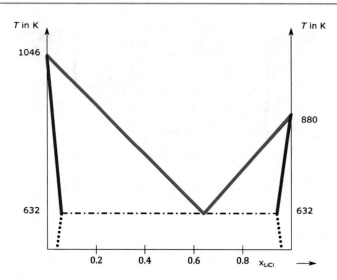

Fig. 8.17 Phase diagram of the system LiCl/KCl

Fig. 8.18 Phase diagram of
the system toluene(A)/water
(B)/acetic acid(C) with
binodal and 4 tie lines

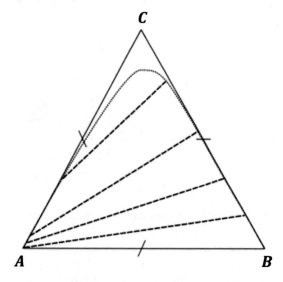

2. Figure 8.18 shows *GIBBS* phase triangle of the system *toluene/water/acetic acid*
   (all data in mass %).

7 kg water, 2 kg acetic acid, and 11 kg toluene are mixed.
   Determine the mass and composition of the "aqueous" and "organic" phases.

Reaction Kinetics

9

9.1 Motivation

If a process has sufficient impetus (i.e., is exergonic), this does not necessarily mean that the process will actually occur. Affinity says nothing about how fast or slow a process takes place. What means do we have to influence on the rate of a process? (The *motivational picture* of this chapter, Fig. 9.1, illustrates the activation energy of an exothermic reaction).

9.2 Does a Process Have an Impetus? (Factsheet: Fig. 9.2)

Thermodynamics is concerned with whether a process—such as the oxyhydrogen reaction here, for example—can take place at all, i.e. whether it has an impetus (affinity) ΔG. If the process is not sufficiently exergonic, we can consider which parameters we can change to increase the impetus. We discussed these parameters—in particular, the effect of temperature and pressure—in the last chapter.

If there is no impetus, the reaction can never occur spontaneously.

If, on the other hand, an impetus exists, the reaction has the "green light" from thermodynamics, so to speak.

But whether the process actually takes place is then a kinetic question: Is reaction rate sufficient?

In this chapter, we will get to know parameters with which we can accelerate the reaction—(here again, temperature and pressure play a major role; in addition, a catalyst can help here). Only when a sufficient impetus and a sufficient reaction rate come together, a process takes place.

© The Author(s), under exclusive license to Springer-Verlag GmbH, DE, part of Springer Nature 2023
J. S. Lauth, *Physical Chemistry in a Nutshell*,
https://doi.org/10.1007/978-3-662-67637-0_9

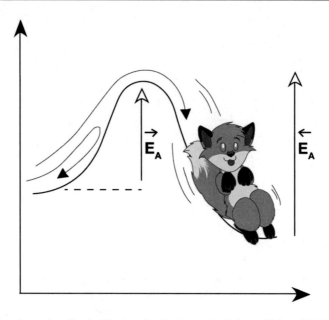

Fig. 9.1 How do we describe the kinetics of a simple reaction? (https://doi.org/10.5446/46040)

Fig. 9.2 Overview of
thermodynamic and kinetic
aspects of the oxyhydrogen
reaction

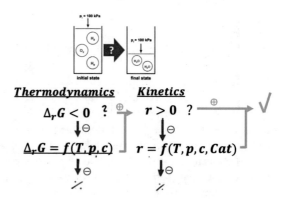

9.3   How Fast Does a Reaction Proceed?

Let us first define the most important parameter of kinetics, the reaction rate.
Figure 9.3 shows the change of the concentration over time of all reactants during
N_2O_5 decomposition. The slope of these curves corresponds to the rate of formation
or decomposition r_i of the respective reactant.

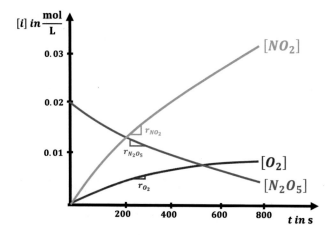

Fig. 9.3   Concentration–time curve of the decay of dinitrogen pentoxide

$$r_i = \frac{d[i]}{dt} \tag{9.1}$$

After 300 s, for example, the decomposition rate of N_2O_5 is $-18\ \mu\text{mol/(L s)}$ and the formation rate of oxygen is $+9\ \mu\text{mol/(L s)}$. If we divide formation or decomposition rates by the respective stoichiometric numbers, we obtain the classical reaction rate r, which is the same for all reactants and products.

$$r = \frac{r_i}{\nu_i} \tag{9.2}$$

In our case

$$r = \frac{r_{NO_2}}{4} = \frac{r_{O_2}}{1} = \frac{r_{N_2O_5}}{-2} = 9\ \frac{\mu\text{mol}}{\text{L s}} \tag{9.3}$$

The unit of reaction rate is always concentration per time, e.g. mol/(L s) or Pa/min.

We see that in this example, the reaction rate is not constant; it is highest at the beginning of the reaction and then decreases asymptotically to zero.

9.4   Which Factors Affect Reaction Rate?

The reaction rate of a reaction

$$A \rightarrow P \tag{9.4}$$

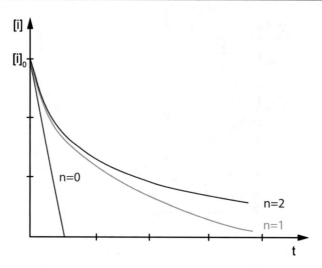

Fig. 9.4 Concentration–time curves (integrated rate laws) of reactions with order 0, 1, and 2

is affected by a number of parameters. These are in particular temperature, the presence of a catalyst, (in the case of heterogeneous reactions) phase boundary, solvent, and concentration of the reactants.

$$r = f(T, \text{catalyst}, \text{phase boundary}, \text{solvent}, [A]) \tag{9.5}$$

The dependence on the concentrations of the reactants is particularly important. Therefore, all other influencing variables are usually combined to form a so-called rate constant k.

$$r = k f([A]) \tag{9.6}$$

The dependence of the reaction rate on the concentrations of the reactants is called the rate law.

Often, but not always, the reaction rate law can be written in the following form:

$$r = k[A]^a \tag{9.7}$$

The exponent of the concentration is called the order. The order of a reaction describes how sensitively the reaction rate is affected by a change in concentration. The order of a reaction very strongly affects the concentration–time curve (see Fig. 9.4).

In general, the order of a reaction cannot be predicted. Experimentally it was found that, e.g., the decay of N_2O_5 is first order, while the decay of NO_2—at first sight a quite similar reaction—is second order.

In a zeroth-order reaction, the reaction rate does not depend on concentration at all. This is the case, for example, with the enzymatic decomposition of ethanol to acetaldehyde. The concentration–time curve is a straight line in this case.

9.5   How Can We Visualize a Reaction on a Mechanical Model? (see Fig. 9.5)

The kinetics of a first-order reaction can be illustrated with the bathtub model: The level of bath tub (1) corresponds to the concentration of the reactant; the level of bath tub (2) corresponds to the concentration of the product. The reaction now proceeds in such a way that a pipette tube is used to transfer water from bath tub (1) to bath tub (2).

We immerse the empty pipette in bath tub (1), close it, remove the (partially) filled pipette and empty its contents into bath tub (2) by opening it.

The size of the pipette corresponds to the rate constant k and the amount of water transferred per unit time corresponds to the reaction rate.

Note that in a first-order reaction, the amount of water transported in the pipette tube is proportional to the water level (concentration) in bath tub (1).

For modeling other reaction orders, the transport vessels or transport methods must be modified.

For example, for a zeroth-order reaction, we must always fill the pipette completely (zeroth-order rate law: $r = k$).

9.6   What Does the Concentration–Time Curve Look Like for a Zeroth-Order Reaction?

Zeroth-order reactions are often found in biochemistry. The rate of these reactions is constant.

The dependence of t concentration with time is also called the integrated rate law. For a zeroth-order reaction this is

$$[A] = [A]_0 - kt \tag{9.8}$$

Plotting $[A]$ against t results in a straight line whose negative slope corresponds to the rate constant.

The reaction in Fig. 10.11, for example, shows a rate constant of

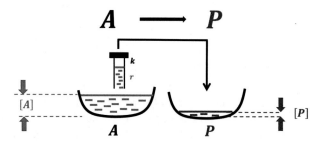

Fig. 9.5 Bathtub model of a first-order reaction

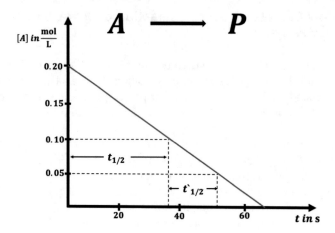

Fig. 9.6 Integrated rate law for a zeroth-order reaction

$$k = -\left(\frac{0.20\,\frac{mol}{L} - 0.050\,\frac{mol}{L}}{0.0\,s - 50\,s}\right) = 0.0030\,\frac{mol}{L\,s} \tag{9.9}$$

The order of a reaction can be specified by analyzing the unit of the rate constant: for a zeroth-order reaction the unit will be mol/(L s).

$$[k] = \frac{mol}{L\,s} \tag{9.10}$$

These integrated velocity laws can be used to calculate yield and reaction times for a chemical reaction.

Yield is defined in this context as

$$yield = \frac{[A]_0 - [A]}{[A]_0} \cdot 100\% \tag{9.11}$$

For the reaction in Fig. 9.6, for example, we can calculate how long it takes for the concentration to drop from an initial 0.2 moL/L to 0.1 mol/L (50% conversion)

$$[A] = [A]_0 - kt \tag{9.12}$$

$$t = \frac{[A]_0 - [A]}{k} = \frac{0.2\,\frac{mol}{L} - 0.1\,\frac{mol}{L}}{0.0030\,\frac{mol}{L\,s}} = 33\,s \tag{9.13}$$

The time it takes for the concentration to drop to half is called the half-life.

In a zeroth-order reaction, half-life is not constant, but becomes shorter and shorter in the course of the reaction. It is related to the rate constant in this way.

$$t_{1/2} = \frac{[A]_0}{2\,k} \qquad (9.14)$$

9.7   What Does the Concentration–Time Curve Look Like for a First-Order Reaction?

Radioactive decay is an example of a first-order process. The rate is proportional to the reactant concentration. If the reactant concentration decreases to half, the rate also decreases to half.

Here we see a compilation of the important formulas—rate law, integrated rate law, half-life—for a first-order reaction (see Fig. 9.7).

The reaction rate is proportional to the concentration in a first-order reaction.

$$r = k \cdot [A]^1 \qquad (9.15)$$

The concentration–time curve for a first-order reaction corresponds mathematically to an exponential function

$$[A] = [A]_0 \cdot e^{-kt} \qquad (9.16)$$

The half-life is constant in a first-order reaction and is directly related to the rate constant.

$$t_{1/2} = \frac{\ln(2)}{k} \qquad (9.17)$$

The unit of the rate constant is $1/s$.

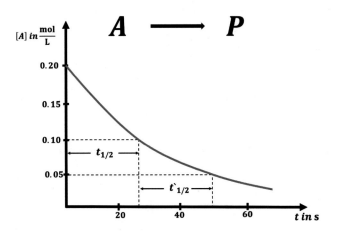

Fig. 9.7 Integrated rate law of a first-order reaction

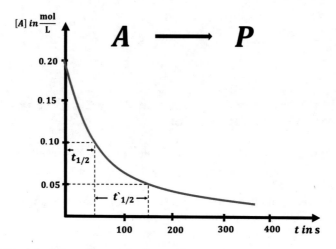

Fig. 9.8 Integrated rate law of a second-order reaction

$$[k] = \frac{1}{s} \tag{9.18}$$

We can linearize this graph by plotting the logarithm of the concentration against time.

The reaction in Fig. 9.8 has a constant half-life of 7.0 s. From this, for example, the rate constant can be determined.

$$k = \frac{\ln(2)}{t_{1/2}} = \frac{\ln(2)}{7.0 \text{ s}} = 0.099 \frac{1}{s} \tag{9.19}$$

A first-order reaction does never go to completion. We can calculate how much reactant is still present after 10 half-lives.

$$[A] = [A]_0 \cdot e^{-kt} \tag{9.20}$$

$$[A] = [A]_0 \cdot e^{-0.099\frac{1}{s} \cdot 70 \text{ s}} \tag{9.21}$$

$$[A] = [A]_0 \cdot 0.00098 \tag{9.22}$$

Only 0.098% of the initial concentration of A is still present. The product yield after 10 half-lives is

$$\text{yield} = \frac{[A]_0 - [A]}{[A]_0} \cdot 100\% \tag{9.23}$$

$$\text{yield} = \frac{[A]_0 - 0.00098\,[A]}{[A]_0} \cdot 100\% = 99.9\% \qquad (9.24)$$

9.8   What Does the Concentration–Time Curve Look Like for a Second-Order Reaction?

The abovementioned decomposition reaction of NO_2 proceeds according to second order. If the concentration of the reactant is halved, the reaction rate is reduced to a quarter.

We can specify the kinetic equations for a second-order reaction as follows:
The reaction rate is proportional to the square of the concentration.

$$r = k \cdot [A]^2 \qquad (9.25)$$

The integrated rate law is a bit more complicated

$$[A] = \frac{[A]_0}{1 + k[A]_0 t} \qquad (9.26)$$

the half-life becomes longer and longer over time

$$t_{1/2} = \frac{1}{k\,[A]_0} \qquad (9.27)$$

and the unit of the rate constant is L/(mol s).

$$[k] = \frac{L}{\text{mol s}} \qquad (9.28)$$

We can linearize the integrated velocity law by plotting the reciprocal of the concentration against time. The slope of the resulting straight line corresponds to the rate constant k.

The reaction in Fig. 9.9 has an initial half-life of 50 s. From this, for example, the rate constant can be determined.

$$k = \frac{1}{t_{1/2}\,[A]_0} = \frac{1}{50\,\text{s} \cdot 0.2\,\frac{\text{mol}}{\text{L}}} = 0.10\,\frac{L}{\text{mol s}} \qquad (9.29)$$

Table 9.1 compiles the important kinetic equations for simple reactions.

In the last line of Fig. 10.2, we also find the corresponding equations for a process $A + B \to P$ with a total order of 2.

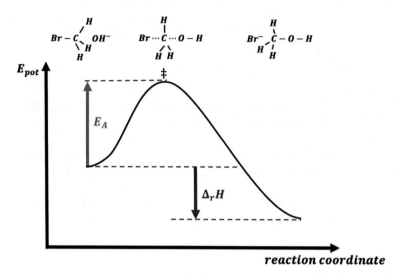

Fig. 9.9 Reaction profile, activated complex (transition state), and activation energy using the example of an SN2 reaction

Table 9.1 Overview "kinetics of simple reactions" (rate laws and integrated rate laws of zeroth-, first-, and second-order reactions)

Reaction	Order	Rate law	$[k]$	Integrated rate law	Half-life
$A \rightarrow P$	0	$r = k$	$\frac{\text{mol}}{\text{L s}}$	$[A] = [A]_0 - kt$	$t_{1/2} = \frac{[A]_0}{2\,k}$
$A \rightarrow P$	1	$r = k[A]$	$\frac{1}{\text{s}}$	$[A] = [A]_0 \cdot e^{-kt}$	$t_{1/2} = \frac{\ln(2)}{k}$
$A \rightarrow P$	2	$r = k\,[A]^2$	$\frac{\text{L}}{\text{mol s}}$	$[A] = \frac{[A]_0}{1+k[A]_0 t}$	$t_{1/2} = \frac{1}{k\,[A]_0}$
$A + B \rightarrow P$	2 (1 + 1)	$r = k[A][B]$	$\frac{\text{L}}{\text{mol s}}$	$kt = \frac{1}{[B]_0 - [A]_0} \ln\left(\frac{[B][A]_0}{[A][B]_0}\right)$	Depending on stoichiometry

9.9   How Does the Potential Energy Change on the Way from the Reactant Molecule to the Product Molecule?

On the one hand, kinetics is a very empirical science that explores rate laws to give us information so that we can run reactions at a controlled rate.

On the other hand, kinetic studies are also basic for understanding a chemical reaction at the molecular level.

What happens microscopically in a chemical reaction? This is specified by the so-called reaction profile.

In every reaction, atoms of the molecules involved rearrange in space. The initial position of the atoms are the reactants (substrates)—the final position of the atoms are the products. On the way from reactants to products (reaction coordinate) energy changes. This is described by the reaction profile.

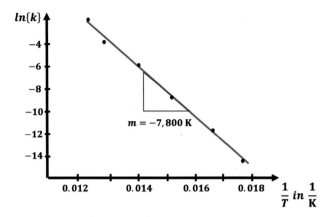

Fig. 9.10 Determining activation energy using Arrhenius plot

Please note that in a kinetic reaction profile—in contrast to the thermodynamic representations of a reaction—the plot actually shows the energy of only very few molecules.

Let us take the SN2 reaction of bromomethane with OH⁻ ions as an example. During the reaction, the atoms of the reactants rearrange in space. This geometric rearrangement is described by the x-axis—the reaction coordinate—in the reaction profile.

The y-axis describes the potential energy. During the reaction, bonds are stretched and split, and other bonds are formed. The potential energy changes during the reaction and passes through a maximum.

The maximum, the most energetic state on the way from the reactants to the products, is called activated complex or transition state and we mark it with a double cross ‡ (or hashtag #).

The energy difference between reactants and transition state is called activation energy.

The transition state and the activation energy now affect the rate of a process to a very decisive extent. The reactant molecules of an exothermic reaction, such as the one shown in Fig. 9.10, also initially require energy to make it "go over the hump," so to speak.

This is usually thermal energy, and it is thus also clear that temperature plays a major role in reaction rate.

9.10 How Does Temperature Affect the Rate of a Reaction?

Table 9.2 presents the kinetic data of the cleavage of cane sugar into glucose and fructose for different temperatures

Table 9.2 Kinetic data of cane sugar inversion at different temperatures

Temperature	Half-life	Rate constant
30.0 ° C	20.0 min	$0.035 \frac{1}{min}$
50.0 ° C	5.0 min	$0.14 \frac{1}{min}$
70.0 ° C	1.0 min	$0.69 \frac{1}{min}$

$$\text{Saccharose} \rightarrow \text{Glucose} + \text{Fructose} \qquad (9.30)$$

The reaction rate constant increases strongly with temperature. VAN'T HOFF's rule states that a temperature increase of 10 °C (10 K, 16 °F) results approximately in a doubling of reaction rate.

ARRHENIUS specified the effect of temperature on k even more mathematically in his famous equation.

$$k = A \cdot e^{-E_A/RT} \qquad (9.31)$$

In ARRHENIUS equation, we find two kinetic parameters, namely the activation energy E_A, which we know from the reaction profile, and the frequency factor A, a limiting rate for infinitely high temperature.

9.11   How Do We Calculate the Kinetic Parameters According to ARRHENIUS?

We can determine the ARRHENIUS *parameters* experimentally if we kinetically measure a reaction at several temperatures and plot the logarithm of the rate constant against the reciprocal of the absolute temperature (linearized form of the ARRHENIUS equation)

$$\ln k = \ln A - \frac{E_A}{R} \frac{1}{T} \qquad (9.32)$$

This plot is called ARRHENIUS plot

From the slope of the resulting straight line, we get the activation energy; from the intercept, we get the frequency factor. If we know the activation energy, we can also use the ARRHENIUS equation to calculate rate constants at any temperature.

$$\ln\left(\frac{k'(T_1)}{k(T_2)}\right) = -\frac{E_A}{R}\left(\frac{1}{T_1} - \frac{1}{T_2}\right) \qquad (9.33)$$

9.12 How Does the Stability of the Transition State Affect Reaction Rate?

The effect of the transition state on reaction rate was quantified in particular by Henry *EYRING*. According to *EYRING*, the rate of a reaction is determined primarily by the difference in stability between the reactants and the transition state, by the free enthalpy of activation $\Delta G^{\#}$.

$$r \sim e^{-\Delta G^{\#}/RT} \tag{9.34}$$

Any parameter that lowers $\Delta G^{\#}$ will accelerate the reaction.

For example, a catalyst acts in such a way that a different, sometimes more complicated, reaction path is taken, but this reduces the activation energy and thus reduces $\Delta G^{\#}$ (see Fig. 9.11).

EYRING's transition state theory can also be used to quantify the influence of the solvent and the effect of ionic strength on reaction rate.

9.13 Summary

The concentration of reactants can affect the reaction rate in different ways.

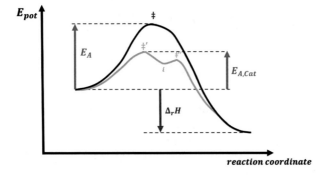

Fig. 9.11 Reaction profile with and without catalyst

Table 9.3 Summary of kinetic data for simple reactions

Reaction	Order	Rate law	$[k]$	Integrated rate law	Half-life
$A \rightarrow P$	0	$r = k$	$\frac{mol}{L\,s}$	$[A] = [A]_0 - kt$	$t_{1/2} = \frac{[A]_0}{2\,k}$
$A \rightarrow P$	1	$r = k[A]$	$\frac{1}{s}$	$[A] = [A]_0 \cdot e^{-kt}$	$t_{1/2} = \frac{\ln(2)}{k}$
$A \rightarrow P$	2	$r = k$ $[A]^2$	$\frac{L}{mol\,s}$	$[A] = \frac{[A]_0}{1+k[A]_0 t}$	$t_{1/2} = \frac{1}{k\,[A]_0}$
$A + B \rightarrow P$	2 (1 + 1)	$r = k[A]$ $[B]$	$\frac{L}{mol\,s}$	$kt = \frac{1}{[B]_0 - [A]_0} \ln\left(\frac{[B]_0[A]_0}{[A][B]_0}\right)$	Depending on stoichiometry

We quantify this by specifying reaction order. Depending on the reaction order, different rate laws, concentration–time curves, and half-lives result (Table 9.3).

Microscopically, a reaction can be illustrated by the reaction profile. Especially the maximum—the transition state—is relevant for the reaction rate. With the help of the ARRHENIUS *equation*

$$k = A \cdot e^{-\frac{E_A}{RT}} \tag{9.35}$$

we can quantify the temperature effect on reaction rate.

We obtain the kinetic parameters activation energy and frequency factor by evaluating the data using the ARRHENIUS *plot*.

Once we know the parameters, we can determine rate constants, yields, and reaction times for arbitrary temperatures. The effects of catalysts, solvents, and ionic strength can be quantified by EYRING'S transition state theory.

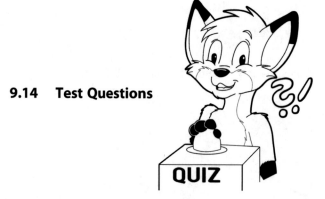

9.14   Test Questions

1. A reaction $A \rightarrow P$ shows a simple second-order kinetics.
   Mark the correct statement(s).
   (a) The reaction rate r has the unit $[r] = \text{mol}/(\text{L*s})$
   (b) The reaction rate constant k has the unit $[k] = 1/s$
   (c) The half-life is constant

    (d) The reaction rate is constant
    (e) The application 1/[A] against *t* results in a straight line

2. The reactant concentration is doubled-
    (a) How does the initial reaction rate and the initial half-life of the reaction change when the reaction shows first-order kinetics?
    (b) How does the initial reaction rate and the initial half-life of the reaction change when the reaction shows second-order kinetics?
3. Reaction A has the activation energy 150 kJ/mol
    Reaction B has the activation energy 100 kJ/mol

    The frequency factors of both reactions are the same and at 50 °C both reactions are equally fast.
    Mark the correct statement(s).

(a) At 70 °C, the reactions are equally fast
(b) At 70 °C, reaction A is faster than reaction B
(c) At 70 °C, reaction B is faster than reaction A
(d) At 30 °C, reaction A is faster than reaction B
(e) At 30 °C, reaction B is faster than reaction A
(f) At 30 °C, the reactions are equally fast

9.15 Exercises

1. Butadiene can dimerize:

$$2\ H_2C=CH-CH=CH_2 \rightarrow \text{Butadien}-\text{Dimer}$$

Table 9.4 Kinetic data for the reaction $A + B \rightarrow P$

$[A]_0$	$[B]_0$	r_0
$0.30\,\frac{mol}{L}$	$0.30\,\frac{mol}{L}$	$0.12\,\frac{mol}{L\,s}$
$0.60\,\frac{mol}{L}$	$0.60\,\frac{mol}{L}$	$0.24\,\frac{mol}{L\,s}$
$0.90\,\frac{mol}{L}$	$0.30\,\frac{mol}{L}$	$0.36\,\frac{mol}{L\,s}$
$0.30\,\frac{mol}{L}$	$0.90\,\frac{mol}{L}$	$0.12\,\frac{mol}{L\,s}$

At 305 °C, pure butadiene with a concentration of 0.0500 mol/L is initially present. The reaction has a rate constant of 9.85 $\frac{mL}{mol\cdot s}$. Calculate

(a) the product yield after 30 minutes
(b) the initial half-life $t_{1/2}$ of the reaction
(c) the initial reaction rate r°

2. The "cane sugar inversion"

$$\text{Saccharose (aq)} + H_2O \rightarrow \text{Fructose(aq)} + \text{Glucose (aq)}$$

proceeds according to first-order kinetics and has a half-life of 10.0 min at 30.0 ° C. At 50.0 °C, the half-life decreases to 2.90 min.
   Calculate the activation energy E_A of the reaction.

3. In the following table, the initial velocities r_0 of the reaction. $A + B \rightarrow P$ for a temperature of 298 K are given (Table 9.4):

The rate law of this reaction is

$$r = k \cdot [A]^a \cdot [B]^b$$

Find the reaction orders a and b and the rate constant k.

Reaction Mechanism

10

10.1 Motivation

Very few reactions proceed in a simple way from reactants to products. Often, there is a whole network of elementary reactions that interlock with each other. These reaction mechanisms mean that the overall kinetics of a reaction can be much more complicated than the cases discussed in the last chapter. (The *motivational picture* of this chapter, Fig. 10.1, illustrates a mechanical analogy for the kinetic and thermo-dynamic control of a parallel reaction).

10.2 What Are the Kinetic Characteristics of a Simple Reaction $A \rightarrow B$?

How does the mechanism affect the rate of a reaction? We will discuss this question in this chapter for the three simplest mechanisms: reversible reaction, consecutive reaction, and parallel reaction.

So far, we have only discussed simple reactions in which the reactants react in <u>one</u> direction to form the products, such as the decomposition of ethylamine to ethylene and ammonia.

$$EtNH_2 \rightarrow CH_2 = CH_2 + NH_3 \tag{10.1}$$

This simple reaction follows a first-order rate law: the rate is proportional to the ethylamine concentration.

$$r = 0.14 \frac{1}{h} \cdot [EtNH_2]^1 \tag{10.2}$$

Accordingly, the integrated rate law is an exponential function.

© The Author(s), under exclusive license to Springer-Verlag GmbH, DE, part of
Springer Nature 2023
J. S. Lauth, *Physical Chemistry in a Nutshell*,
https://doi.org/10.1007/978-3-662-67637-0_10

Fig. 10.1 How do we
describe the kinetics of more
complex reactions? (https://
doi.org/10.5446/46043)

$$[EtNH_2] = [EtNH_2]_0 \cdot e^{-0.14 \frac{1}{h} \cdot t} \tag{10.3}$$

The rate constant depends on temperature in the way specified by ARRHENIUS. If we know the ARRHENIUS parameters activation energy and frequency factor, we can determine k for arbitrary temperatures.

$$k = 8.12 \cdot 10^{10} \frac{1}{s} \cdot e^{-\frac{176.4 \frac{kJ}{mol}}{RT}} \tag{10.4}$$

10.3   What Elementary Reactions Does a Reaction Consist of?

In many cases, however, a reactant does not react in one step to form the product. For the cleavage of acetaldehyde, for example, a reaction order of 1.5 has been determined experimentally.

$$CH_3CHO \rightarrow CH_4 + CO \tag{10.5}$$

$$r = k \cdot [CH_3CHO]^{1.5} \tag{10.6}$$

This rate law can only be understood under the prerequisite, that this reaction does not occur in one step, but involves three elementary reactions.

$$CH_3CHO \overset{k}{\rightarrow} CH_3 \cdot + CHO \quad (I) \tag{10.7}$$

$$CH_3 \cdot + CH_3CHO \overset{k'}{\rightarrow} CH_4 \cdot + CO + CH_3 \cdot \quad (II) \tag{10.8}$$

$$CH_3 \cdot + CH_3 \cdot \overset{k''}{\rightarrow} C_2H_6 \quad (III) \tag{10.9}$$

These three elementary reactions make up the mechanism of the reaction. Each elementary reaction has a simple kinetics—as described in the last chapter

$$r_1 = k \, [CH_3CHO] \tag{10.10}$$

$$r_2 = k'[CH_3\cdot][CH_3CHO] \tag{10.11}$$

$$r_3 = k''[CH_3\cdot]^2 \tag{10.12}$$

but the interlocking of the reactions makes the overall kinetics a bit more complicated.

Elementary reactions are those reactions which occur just as specified at the molecular level.

For elementary reactions, we can even predict the reaction order: unimolecular reactions follow a first-order kinetics and bimolecular reactions follow a second-order kinetics.

The terms *unimolecular* and *bimolecular* describe how many reactant molecules make up the transition state.

Thus, the kinetics of acetaldehyde decomposition can be understood if we postulate a mechanism of three elementary reactions I, II, and III. Elementary reaction I is unimolecular and first order; elementary reactions II and III are bimolecular and second order.

10.4 What Mechanisms Can We Combine Using Two Elementary Reactions?

We will discuss in this chapter mechanisms built from two elementary reactions.

In total, we can construct three mechanisms from two elementary reactions, which we will discuss in more detail below. These are the reversible reaction mechanism, the consecutive reaction mechanism, and the parallel reaction mechanism (see Fig. 10.2).

Fig. 10.2 Reaction
mechanisms with two
elementary reactions:
Reversible reaction,
consecutive reaction, and
parallel reaction

$$A \rightarrow B$$
$$B \rightarrow A$$

$$A \rightarrow B$$
$$B \rightarrow C$$

$$A \rightarrow B$$
$$A \rightarrow C$$

10.5   How Do We Describe the Mechanism of a Reversible Reaction?

An example of an reversible reaction is the conversion of α-D-glucose into β-D-glucose—the so-called mutarotation.

$$\alpha - D - Glucose \underset{\overleftarrow{k}}{\overset{\overrightarrow{k}}{\rightleftarrows}} \beta - D - Glucose \tag{10.13}$$

In the so-called forward reaction α-glucose converts to β-glucose. This elementary reaction is unimolecular and consequently proceeds according to a first-order rate law.

$$\overrightarrow{r} = - \left(\frac{d[\alpha]}{dt} \right)_{\rightarrow} \tag{10.14}$$

$$\overrightarrow{r} = \overrightarrow{k} \cdot [\alpha] \tag{10.15}$$

In the reverse reaction β-glucose converts to α-glucose also in a unimolecular elementary reaction. However, the rate constant is different from the forward reaction.

$$\overleftarrow{r} = \left(\frac{d[\alpha]}{dt} \right)_{\leftarrow} \tag{10.16}$$

$$\overleftarrow{r} = \overleftarrow{k} \cdot [\beta] \tag{10.17}$$

10.6   How Can We Represent a Reversible Reaction in a Model?

Using our bathtub model, we can picture a reversible reaction as not only transferring water from the reactant bathtub to the product bathtub with a pipette tube ($\overrightarrow{k}$), but also, conversely, transferring water from the product bathtub back to the reactant bathtub with another pipette tube ($\overleftarrow{k}$) (see Fig. 10.3).

10.7   What Do the Concentration–Time Curves Look Like in a Reversible Reaction?

For a complete description of the kinetics of the reversible reaction

$$A \underset{\overleftarrow{k}}{\overset{\overrightarrow{k}}{\rightleftarrows}} B \tag{10.18}$$

we have to balance the elementary reactions. In balance terms, the reverse reaction represents a source for reactant A; the reverse reaction, on the other hand, is a sink for reactant A.

The total change in concentration of A is thus given by

$$\frac{d[A]}{dt} = \overleftarrow{k}\,[B] - \overrightarrow{k} \cdot [A] \tag{10.19}$$

This is the rate law of an reversible reaction, which we can integrate and get the concentration–time curves from.

$$[A] = \frac{[A]_0}{\overrightarrow{k} + \overleftarrow{k}} \left(\overleftarrow{k} + \overrightarrow{k} \cdot e^{-\left(\overrightarrow{k} + \overleftarrow{k}\right)t} \right) \tag{10.20}$$

The concentration of A decreases exponentially, but not to 0, but to an equilibrium value $[A]_{eq}$ (see Fig. 10.4)

In the same way, the concentration of B increases starting from 0 to an equilibrium value $[B]_{eq}$.

Fig. 10.3 *Bathtub model of a reversible reaction*

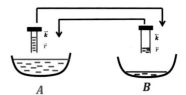

A B

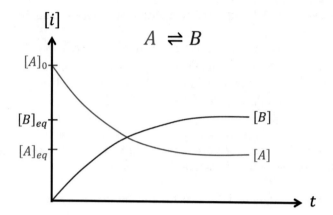

Fig. 10.4 Integrated rate law of a reversible reaction $A \rightleftharpoons B$

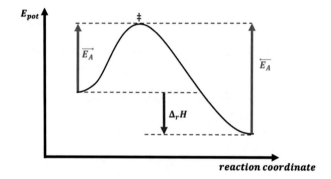

Fig. 10.5 Reaction profile of a reversible reaction

10.8 What Is the Relationship Between Kinetics and Thermodynamics in a Reversible Reaction? (see Fig. 10.5)

In the reaction profile, we now need both an activation energy $\overrightarrow{E_A}$ of the forward reaction and the activation energy $\overleftarrow{E_A}$ of the reverse reaction. The difference of these activation energies is the reaction enthalpy $\Delta_R H$ a thermodynamic quantity.

$$\overrightarrow{E_A} - \overleftarrow{E_A} = \Delta_R H \tag{10.21}$$

In dynamic equilibrium, the rate of the forward reaction is the same as the rate of the reverse reaction.

$$\overleftarrow{r} = \overrightarrow{r} \tag{10.22}$$

If we formulate the velocity laws accordingly, we get the law of mass action:

$$\overleftarrow{k}\,[B]_{eq} = \overrightarrow{k} \cdot [A]_{eq} \qquad (10.23)$$

$$K_{eq} = \frac{[B]_{eq}}{[A]_{eq}} = \frac{\overrightarrow{k_1}}{\overleftarrow{k_1}} \qquad (10.24)$$

10.9   How Do We Describe the Mechanism of a Consecutive Reaction?

In the radioactive decay series of uranium, radium decays into radon and this further decays into polonium (see Fig. 10.6).

This is an example of a consecutive reaction. Radon is the intermediate product that forms in the first reaction (formation reaction) and decays in the second reaction (decay reaction).

More generally formulated:

$$
\begin{array}{ccc}
k & k' \\
A \rightarrow B \rightarrow C
\end{array} \qquad (10.26)
$$

The formation reaction is the source for intermediate B and the decay reaction is the sink for intermediate B.

10.10   How Can We Represent a Consecutive Reaction in a Model?

Using our bathtub model, we now have to consider three tubs A, B, and C. We use pipette tube (k) to transfer water from A to B and another pipette tube (k') to transfer water from B to C. The respective size of the transport vessel corresponds to the rate constant (see Fig. 10.7).

$$
{}^{226}_{88}Ra \quad \overset{\alpha}{\underset{t_{1/2}\,=\,1{,}622\ \text{a}}{\longrightarrow}} \quad {}^{222}_{86}Rn \quad \overset{\alpha}{\underset{t_{1/2}\,=\,3.8\ \text{d}}{\longrightarrow}} \quad {}^{218}_{84}Po
$$

Fig. 10.6 Part of the uranium–radium decay series

Fig. 10.7 Bathtub model of a consecutive reaction

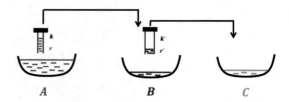

10.11   What Do the Concentration–Time Curves Look Like in a Consecutive Reaction?

For the overall kinetics, we have to consider the formation reaction of the intermediate product

$$r_1 = \left(\frac{d[B]}{dt}\right)_I \tag{10.27}$$

$$r_1 = k \cdot [A] \tag{10.28}$$

and the decomposition reactions of the intermediate product

$$r_2 = - \left(\frac{d[B]}{dt}\right)_{II} \tag{10.29}$$

$$r_2 = k'[B] \tag{10.30}$$

Balancing source and sink of the intermediate, we obtain the rate law.

$$\frac{d[B]}{dt} = k \cdot [A] - k' \cdot [B] \tag{10.31}$$

After integrating, we end up with the concentration–time curves (see Fig. 10.8).

$$[A] = [A]_0 \cdot e^{-k\,t} \tag{10.32}$$

$$[B] = \frac{k}{k'-k} \cdot [A]_0 \cdot \left(e^{-kt} - e^{-k't}\right) \tag{10.33}$$

$$[C] = [A]_0 \cdot \left(1 - \frac{k'}{k'-k} e^{-kt} + \frac{k}{k'-k} e^{-k't}\right) \tag{10.34}$$

Of particular interest is the concentration–time curve of intermediate B, which shows a maximum.

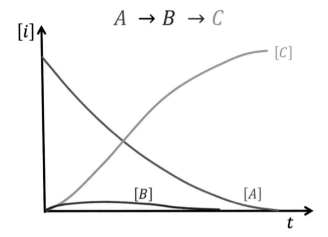

Fig. 10.8 Concentration–time curves for a consecutive reaction $A \rightarrow B \rightarrow C$ with reactive intermediate B

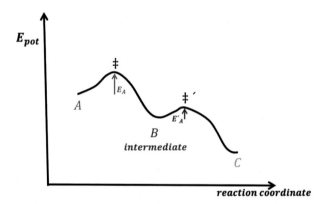

Fig. 10.9 Reaction profile for a consecutive reaction with reactive intermediate

10.12 Is the Intermediate Stable or Reactive?

If the intermediate is stable, a pronounced maximum appears in the concentration-time curve. In the case of a reactive intermediate, the maximum will be very flat and hardly discernible.

In the latter important case, the steady-state principle applies.

In the reaction profile of a consecutive reaction, we find two maxima and a minimum. The maxima correspond to the transition states and the minimum corresponds to the intermediate product (see Fig. 10.9).

Accordingly, there are also two activation energies. If the rate constants of the formation and decay reactions are very different, the slowest reaction is the rate-determining step for consecutive reactions.

If the decomposition reaction has a significantly larger rate constant than the formation reaction, the intermediate is referred to as a reactive intermediate. Here, BODENSTEIN'S steady-state principle, already briefly mentioned above, applies: The intermediate has a very low concentration, which practically does not change in time.

$$\frac{d[B]}{dt} = k[A] - k'[B] \approx 0 \qquad\qquad (10.35)$$

$$[B] \approx 0 \qquad\qquad (10.36)$$

10.13 How Do We Describe the Mechanism of a Parallel Reaction?

Butadiene can add hydrogen bromide, either to the 1,2- or to the 1,4-product (see Fig. 10.10).

This is an example of a parallel reaction. The reactant A can react "to the right" to the product B, on the one hand, and "to the left" to the product C, on the other hand.

$$\overset{k'}{C \leftarrow} \overset{k}{A \rightarrow B} \qquad\qquad (10.37)$$

10.14 How Can We Represent a Parallel Reaction in a Model?

Using our bathtub model, we now transfer water from one and the same bathtub A to bathtubs B and C using two different pipette tubes (see Fig. 10.11).

$$H_2C = CH - CH = CH_2 \quad + \quad HBr \quad \rightarrow \quad H_3C - CHBr - CH = CH_2$$

$$H_2C = CH - CH = CH_2 \quad + \quad HBr \quad \rightarrow \quad H_3C - CH = CH = CH_2Br$$

Fig. 10.10 Hydrobromination of butadiene as an example of a parallel reaction (competitive reaction)

Fig. 10.11 Bathtub model of a parallel reaction

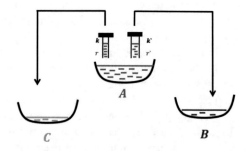

10.15   What Do the Concentration–Time Curves Look Like for a Parallel Reaction?

As before, we obtain the kinetics of the overall reaction by balancing the two elementary reactions. In this case, both reactions act as sinks for the reactant A.

$$r_1 = -\left(\frac{d[A]}{dt}\right)_I = k \cdot [A] \tag{10.38}$$

$$r_2 = -\left(\frac{d[A]}{dt}\right)_{II} = k' \cdot [A] \tag{10.39}$$

Therefore, there are two negative terms in the rate law.

$$\frac{d[A]}{dt} = -k[A] - k' \cdot [A] \tag{10.40}$$

By integrating this law, we obtain the concentration–time curves (Fig. 10.12).

$$[A] = [A]_0 \cdot e^{-(k'+k)\,t} \tag{10.41}$$

$$[B] = \frac{k}{k'+k} \cdot [A]_0 \cdot \left(1 - e^{-(k'+k)\,t}\right) \tag{10.42}$$

$$[C] = \frac{k'}{k'+k} \cdot [A]_0 \cdot \left(1 - e^{-(k'+k)\,t}\right) \tag{10.43}$$

It is interesting that the quotient of the two product concentrations [B] and [C] is constant—the so-called WEGSCHEIDER principle of constant selectivity.

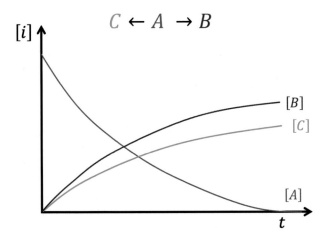

Fig. 10.12 Concentration–time curves of a parallel reaction (B is the "kinetic" product)

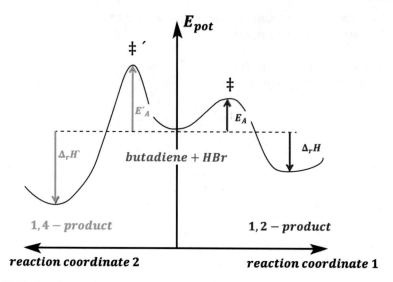

Fig. 10.13 Reaction profile of a parallel reaction (hydrobromination of butadiene); kinetic and thermodynamic control

$$\frac{[B]}{[C]} = \frac{k}{k'} \tag{10.44}$$

In the reaction profile of a parallel reaction the reactant is located in the center and starting from it two reaction coordinates proceed to the right and left to the products (see Fig. 10.13).

10.16   Does the Kinetic or the Thermodynamic Product Dominate?

If the activation energies and thus the rate constants are significantly different in a parallel reaction, the fastest reaction determines the kinetics of the overall process. However, this is only true if the reaction is kinetically controlled, i.e. if we have a shortage of thermal energy and time.

In our example, the 1,2-addition product is the kinetic product: it is formed faster and at low temperature and short reaction time it is the main product.

At high temperatures and long reaction times, different activation energies are no longer this relevant. Then thermodynamic stability decides which will be the main product.

In our example, the 1,4 product is the thermodynamic product—it has lower energy, higher thermodynamic stability and is formed as the main product at high temperatures and long reaction times. Selectivity then corresponds to the thermodynamic stabilities and thus to the equilibrium constants of the reactions.

$$\frac{[B]}{[C]} = \frac{K_{eq}}{K'_{eq}} \tag{10.45}$$

The first situation $\left(\frac{[B]}{[C]} = \frac{k}{k'}\right)$ is labeled "kinetic control" and the second situation $\left(\frac{[B]}{[C]} = \frac{K_{eq}}{K'_{eq}}\right)$ is labeled "thermodynamic control."

The rule of thumb is "CSCK": Cold reaction temperature, short reaction time and catalyst for kinetic control.

$$T \downarrow \, ; t \downarrow \, ; \text{Catalyst} \tag{10.46}$$

10.17 Summary

In reversible reactions, the rates of the forward and reverse reactions are equal at equilibrium.

We can derive the law of mass action kinetically.

$$K_{eq} = \frac{[B]_{eq}}{[A]_{eq}} = \frac{\overrightarrow{k_1}}{\overleftarrow{k_1}} \tag{10.47}$$

Furthermore, we see from the reaction profile that there is a relationship between activation energies and enthalpy of reaction.

$$\overrightarrow{E_A} - \overleftarrow{E_A} = \Delta_R H \tag{10.48}$$

For consecutive reactions with reactive intermediate, the steady-state principle applies: the concentration of this intermediate is approximately 0 and constant in time.

$$\frac{d[B]}{dt} = k_1[A] - k'_1 \cdot [B] \approx 0 \tag{10.49}$$

In parallel reactions, the reaction path correlated with the lower activation energy is primarily taken—provided the reaction is kinetically controlled.

$$\frac{[B]}{[C]} = \frac{k_1}{k_1'}$$

(10.50)

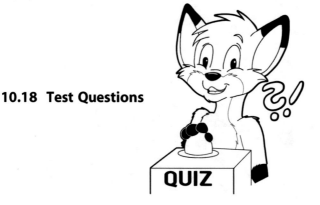

10.18 Test Questions

1. At equilibrium...
    (a) the rates of the forward reaction and the reverse reaction are equal
    (b) the rate constants of the forward reaction and reverse reaction are equal
    (c) the activation energies of the forward reaction and reverse reaction are equal
    (d) Formation rate and decay rate of the product are equal

2. What is the rate-determining step in a...
    (a) Consecutive reaction
    (b) Parallel reaction

3. What statement(s) does the BODENSTEIN steady-state principle make?
    (a) The concentration of a reactive intermediate is very low
    (b) The concentration of a stable intermediate is very low
    (c) Formation rate and decay rate of an intermediate are equal
    (d) The concentration of a reactive intermediate is constant in time

4. Mark the correct statement(s).
    (a) For kinetically controlled reaction, the temperature should be as high as possible
    (b) For thermodynamically controlled reaction, the temperature should be as high as possible
    (c) For kinetically controlled reactions, a catalyst should be used

10.19 Exercises

1. Consider a reversible reaction

$$A \underset{\overleftarrow{k}}{\overset{\overrightarrow{k}}{\rightleftarrows}} B$$

The forward reaction has the activation energy $\overrightarrow{E_A} = 11.9 \frac{kJ}{mol}$ and at 24.0 °C the rate constant $\overrightarrow{k}\,(24.0\,°C) = 11.9 \frac{1}{h}$

The reverse reaction has the activation energy $\overleftarrow{E_A} = 19.4 \frac{kJ}{mol}$ and at 24.0 °C the rate constant $\overleftarrow{k}\,(24.0\,°C) = 2.60 \frac{1}{h}$

(a) Calculate the enthalpy of reaction $\Delta_r H$

(b) Calculate the constant of the forward reaction $\overrightarrow{k}\,(40.5\,°C)$ at 40.5 °C

(c) Calculate the equilibrium constant $K_{eq}(24.0\,°\,C)$ of the reaction.

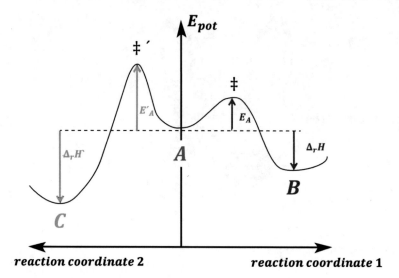

Fig. 10.14 Reaction profile of a parallel reaction

2. A parallel reaction, starting from reactant A, has the following reaction profile (see Fig. 10.14)

What are the conditions for B to be the main product? What are the conditions for C to be the main product?

Conductivity

11

11.1 Motivation

Electrochemical phenomena are ubiquitous in everyday life and technology: a battery converts chemical energy into electrical energy; a pH sensor turns chemical information into electrical voltage, etc. The basis of all electrochemical phenomena is the transport of charges in various conductive materials. (The *motivational picture* of this chapter, Fig. 11.1, illustrates the flow of electric current in an electrolyte.).

11.2 How Does Charge Transport Work in an Electronic Conductor?

In charge transport, we distinguish between electronic and ionic conductors ("first and second class conductors," "first and second type conductors").

Metals such as copper or silver are electronic conductors. Here, charge transport occurs exclusively through electrons.

Figure 11.2 shows the electron gas model of silver: The positively charged silver ions form an immobile crystal lattice; the electrons move freely between the silver ions like a gas. Each electron has the negative elementary charge.

$$Q = -e = 1.6 \cdot 10^{-19} \, \text{As} \qquad (11.1)$$

One mole of electrons has the charge 96,485 C—this is FARADAY'S constant F.

$$e \cdot N_A = 96485 \, \frac{\text{As}}{\text{mol}} = F \qquad (11.2)$$

In addition to metals and semimetals, intrinsically conductive polymers (ICPs) are also electronic conductors.

© The Author(s), under exclusive license to Springer-Verlag GmbH, DE, part of Springer Nature 2023
J. S. Lauth, *Physical Chemistry in a Nutshell*,
https://doi.org/10.1007/978-3-662-67637-0_11

Fig. 11.1 How do ions move in electrolytes? (https://doi.org/10.5446/46041)

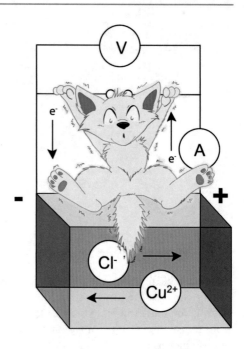

Fig. 11.2 Electron gas model of an electronic conductor (silver)

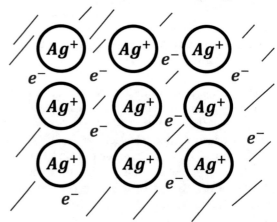

11.3 How Does Charge Transport Work in an Ionic Conductor? (see Fig. 11.3)

In electrolytes, charge transport occurs via ions. Cations carry a positive charge

$$Q^+ = z^+ e \tag{11.3}$$

Anions carry a negative charge

Fig. 11.3 Charge transport in
a second-class conductor

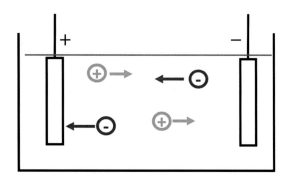

$$Q^- = z^- e \tag{11.4}$$

11.4   What Does the Structure of an Electrolyte Look like?

We can prepare an electrolyte, for example, by dissolving a salt in water.

$$K_{\nu^+} A_{\nu^-} \xrightarrow{\alpha} \nu^+ K^{z^+} + \nu^- A^{z^-} \tag{11.5}$$

ν^+ and ν^- are the decay numbers, z^+ and z^- are the charge numbers of the ions. When an electrolyte is formed, there are always as many positive charges as negative charges. Their number is quantified by the electrochemical coefficient n_e.

$$n_e = \nu^+ z^+ = |\nu^- z^-| \tag{11.6}$$

For example, 1 mole of positive charges and 1 mole of negative charges are generated from one mole of common salt:

$$NaCl \rightarrow Na^+ + Cl^- \tag{11.7}$$

$$n_e = 1(+1) = |1(-1)| = 1 \tag{11.8}$$

The ions are hydrated in the electrolyte; during electrical conduction, the hydrate shell will stay at the ion and migrate along with them (see Fig. 11.4).

Fig. 11.4 Size comparison of
a bare and a hydrated
copper ion

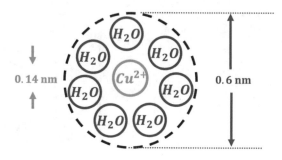

Fig. 11.5 Ionic cloud around
a hydrated copper ion

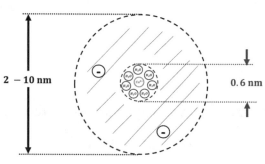

11.5   What Is the Effective Concentration (Activity) of an Electrolyte?

Due to the strong electrical interactions of the ions, we may only assume in extremely dilute solutions that the ions do not interact with each other ("ideal electrolytes"). In real electrolytes, the interactions usually have to be taken into account. The situation in an electrolyte was described by *DEBYE* and *HÜCKEL* in the theory named after them: The ions are shielded in the electrolyte by an oppositely charged ion cloud ("ionic atmosphere") (see Fig. 11.5).

As a result, their effective concentration—the activity a—is different from their weigh-in concentration c.

$$a_{\pm} = f_{\pm} \cdot c_{\pm} \tag{11.9}$$

The *DEBYE–HÜCKEL* limiting law (valid up to a concentration of approx. 0.01 mol/L) can be used to calculate the activity coefficient f.

$$\log f_{\pm} = 0.509 \, z_{+} \, z_{-} \sqrt{\frac{I}{\text{mol/L}}} \tag{11.10}$$

I is the ionic strength, a kind of extended concentration value.

$$I = \frac{1}{2}\sum_i z_i^2 \cdot c_i \tag{11.11}$$

In a saline solution, the concentration and ionic strength are the same; a saline solution of the weigh-in concentration 0.5 mol/L

$$I = c = 0.5 \frac{\text{mol}}{\text{L}} \tag{11.12}$$

has an activity coefficient of

$$f_\pm = 0.7 \tag{11.13}$$

thus has an effective concentration of only

$$a_\pm = 0.35 \frac{\text{mol}}{\text{L}} \tag{11.14}$$

In all thermodynamic equations in which the concentration occurs, e.g. in the law of mass action, in the ion product, or in the pH value, the activity rather than the concentration must be used for accurate calculations.

11.6   How Do We Measure the Electrical Conductivity of an Electrolyte?

Using a classical electrical circuit with voltmeter and ammeter, we can measure the conductivity of an electrolyte (see Fig. 11.6).

If we put a 0.5 molar saline solution into this experimental setup and apply 5 volts of voltage, 0.25 A will flow. Applying *Ohm*'s law, we calculate a resistance R of 20 Ω or a conductance (reciprocal of R) of 50 mS, respectively.

Fig. 11.6 Test setup for measuring the electrical conductivity of a saline solution

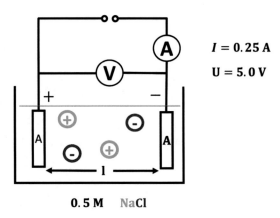

$I = 0.25\,\text{A}$

$U = 5.0\,\text{V}$

0.5 M   NaCl

$$R = \frac{U}{I} = \frac{5.0 \text{ V}}{0.25 \text{ A}} = 20 \ \Omega; \ \frac{1}{R} = \frac{1}{20 \ \Omega} = 50 \text{ mS} \qquad (11.15)$$

Normalizing conductance to the size of the measuring cell, we define electrical conductivity κ.

$$\kappa = \frac{1}{R} \cdot \frac{l}{A} = \frac{1}{20 \ \Omega} \cdot \frac{1 \text{ m}}{0.01 \text{ m}^2} = 5.0 \frac{\text{S}}{\text{m}} \qquad (11.16)$$

The electrical conductivity of our saline solution is about one million times greater than the conductivity of purest water.

$$\kappa(H_2O) = 5.5 \frac{\mu S}{m} \qquad (11.17)$$

Purest water contains at room temperature 0.1 nmol/L protons and 0.1 nmol/L hydroxide ions due to autoprotolysis, which cause this conductivity.

The conductivity of a medium is a measure of how many charge carriers are present and how mobile they are. Naturally, the conductivity is highest in metals (especially copper).

$$\kappa(Cu) = 58 \frac{MS}{m} \qquad (11.18)$$

In the case of electrolytes, the conductivity depends strongly on the concentration.

A 0.05 molar saline solution contains only one tenth of the ions of the solution just discussed and conducts about a factor of 10 worse.

$$\kappa(0.5 \text{ M NaCl}) = 5.0 \ \frac{S}{m} \qquad (11.19)$$

$$\kappa(0.05 \text{ M NaCl}) = 0.6 \ \frac{S}{m} \qquad (11.20)$$

11.7 How Do We Obtain the Molar Conductivity of an Electrolyte from the Specific Conductivity?

We can normalize the conductivity of an electrolyte to one mole of electrolyte by dividing the specific conductivity κ by the concentration c. We then obtain the molar conductivity Λ.

$$\Lambda = \frac{\kappa}{c} \qquad (11.21)$$

For our saline solution it is

$$\Lambda = \frac{5.0 \frac{S}{m}}{500 \frac{mol}{m^3}} = 10 \ \frac{mS \ m^2}{mol} \tag{11.22}$$

The molar conductivity is a measure of how well 1 mol of electrolyte conducts.

Another possibility is to normalize conductivity of an electrolyte to one mole of charge. If we divide the molar conductivity Λ by the electrochemical coefficient, we get the equivalent conductivity Λ_e.

$$\Lambda_e = \frac{\Lambda}{n_e} \tag{11.23}$$

For 1–1 electrolytes such as saline, molar conductivity and equivalent conductivity agree.

11.8   How Does the Molar Conductivity of an Electrolyte Change when Diluted?

If we dissolve (in a thought experiment) 1 mol of electrolyte in infinite solvent, we could measure in this ideal electrolyte the so-called limiting molar conductivity Λ_∞ (see Fig. 11.7).

If the electrolyte behaved ideally even at finite concentrations, we would always measure this limiting molar conductivity; the molar conductivity of an ideal electrolyte would be constant.

$$\Lambda = \Lambda_\infty = \text{const} \tag{11.24}$$

In real electrolytes, molar conductivity increases with decreasing concentration because the mobility of the ions increases. The concentration dependence of molar conductivity ($\Lambda = f(c)$) and thus the approach to limiting molar conductivity is different for strong and for weak electrolytes.

For strong electrolytes, KOHLRAUSCH'S square root law applies,

Fig. 11.7 Molar conductivity of ideal, strong, and weak electrolytes

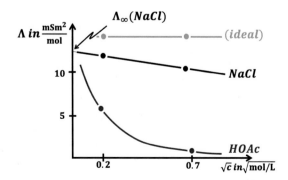

$$\Lambda = \Lambda_\infty - K_K \sqrt{c}. \tag{11.25}$$

The limiting molar conductivity of the strong electrolyte common salt is

$$\Lambda_\infty(\text{NaCl}) = 12.64 \frac{\text{mS m}^2}{\text{mol}} \tag{11.26}$$

We can obtain this value by measuring several saline solutions of different concentrations and extrapolating according to KOHLRAUSCH'S square root law (see Fig. 12.5).

For weak electrolytes, the relationship between molar conductivity and limiting molar conductivity looks relatively simple at first.

$$\Lambda = \alpha \cdot \Lambda_\infty \tag{11.27}$$

However, the degree of dissociation α itself will depend on concentration; it is related to the acid dissociation constant K_a via OSTWALD'S dilution law.

$$K_a = \frac{c \cdot \alpha^2}{1 - \alpha} \tag{11.28}$$

Combining the equations results in a more complicated function, which we may use to experimentally determine the limiting molar conductivity of acetic acid (see lab experiments).

$$\frac{1}{\Lambda} = \frac{1}{\Lambda_\infty} + \frac{\Lambda_c}{K_a(\Lambda_\infty)^2} \tag{11.29}$$

11.9   How Can We Calculate the Limiting Conductivity of an Electrolyte?

The limiting molar conductivity of an electrolyte has two contributions: the limiting ionic conductivity of the cations $\lambda_{+\infty}$ and the limiting ionic conductivity of the anions $\lambda_{-\infty}$

$$\Lambda_\infty = \nu_+ \lambda_{+\infty} + \nu_- \lambda_{-\infty} \tag{11.30}$$

This equation also goes back to KOHLRAUSCH, it is the law of independent ion migration. From tabulated ionic conductivities, this equation can be used to calculate limiting molar conductivities for any electrolytes.

The ionic conductivity λ_∞ of an ion mainly depends on its hydrodynamic radius, i.e. on its size including the hydrate shell (see Table 11.1).

Potassium ions, for example, do conduct better than sodium ions because potassium ions have a smaller hydrate shell. In water, H^+ ions and OH^- ions conduct best by far. This is because these ions use a special conductivity mechanism—the GROTTHUß mechanism.

Table 11.1 Limiting molar conductivities of some ions

$\lambda_\infty(K^+) = 7.35 \frac{mS\ m^2}{mol}$
$\lambda_\infty(Na^+) = 5.01 \frac{mS\ m^2}{mol}$
$\lambda_\infty(H^+) = 35.0 \frac{mS\ m^2}{mol}$
$\lambda_\infty(Cl^-) = 7.63 \frac{mS\ m^2}{mol}$
$\lambda_\infty(OH^-) = 19.9 \frac{mS\ m^2}{mol}$
$\lambda_\infty(Acetat^-) = 4.09 \frac{mS\ m^2}{mol}$

11.10 What Is the Contribution of the Cation or Anion to Electrical Conductivity?

As a rule, anions and cations make different contributions to conductivity. This may be quantified using the so-called ion transport numbers t_+ and t_-.

$$t_+ = \frac{\nu_+ \lambda_+}{\Lambda} \tag{11.31}$$

$$t_- = \frac{\nu_- \lambda_-}{\Lambda} \tag{11.32}$$

The chloride ion, for example, conducts much better than the sodium ion; its transport number (or transference number) t_+ in saline solution is

$$t_+ = \frac{5.01}{12.64} = 0.40 \tag{11.33}$$

This means: 40% of the charge transport in saline solution is accomplished by cations, and 60% is accomplished by anions.

11.11 How Fast Does an Ion Move in the Electric Field? (see Fig. 11.8)

Microscopically, conductivity can be discussed using Fig. 13.13: the ions are in an electric field. The field force

$$F_{el} = z\, e\, E \tag{11.34}$$

initially accelerates the cations toward the cathode until the field force is compensated by an oppositely directed frictional force.

$$F_{Stokes} = 6\, \eta\, \pi\, r\, v \tag{11.35}$$

The ions finally will have a constant drift velocity.

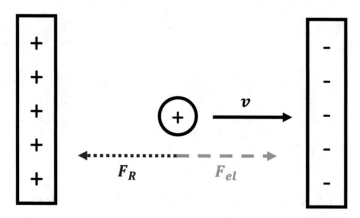

Fig. 11.8 Equilibrium of forces on an ion in the electric field

$$v_+ = \frac{z_+ e}{6\,\eta\,\pi\,r} \cdot E \tag{11.36}$$

The drift velocity depends on the electric field strength E and an ion-specific factor called electrical mobility u

$$u_+ = \frac{z_+ e}{6\,\eta\,\pi\,r} \tag{11.37}$$

The electrical mobility u_+ (resp. u_-) can be easily determined using ionic molar conductivity λ_+ (resp. λ_-) and *Faraday's constant*.

$$u_+ = \frac{\lambda_+}{F} \tag{11.38}$$

In saline solution, the chloride ion has the greater mobility; thus, chloride ions will show a higher drift velocity than sodium ions in this solution.

11.12 Summary

For the complete description of an electrolyte

$$K_{\nu_+}A_{\nu_-} \xrightarrow{\alpha} \nu_+ K^{z+} + \nu_- A^{z-} \tag{11.39}$$

we need its electrochemical coefficient,

$$n_e = \nu^+ z^+ = |\nu^- z^-| \tag{11.40}$$

its ionic strength

$$I = \frac{1}{2}\sum_i z_i^2 \cdot c_i \tag{11.41}$$

and its activity or its activity coefficients according to DEBYE–HÜCKEL

$$\log f_\pm = 0.509 \, z_+ \, z_- \sqrt{\frac{I}{\frac{\text{mol}}{\text{L}}}} \tag{11.42}$$

In addition to the classic electrical conductivity κ

$$\kappa = \frac{1}{R}\frac{l}{A} \tag{11.43}$$

the molar conductivity Λ and molar limiting conductivity Λ_∞ are introduced for electrolytes.

$$\Lambda = \frac{\kappa}{c} \tag{11.44}$$

The molar limiting conductivity is composed of the anion and cation proportion.

$$\Lambda_\infty = \nu^+ \lambda_\infty^+ + \nu^- \lambda_\infty^- \tag{11.45}$$

This proportion can be quantified by means of the transport numbers.

$$t_+ = \frac{\nu^+ \lambda_+}{\Lambda} \tag{11.46}$$

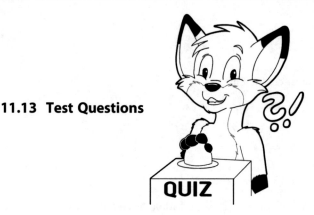

11.13 Test Questions

1. Using the table, rank the molar limiting conductivities of the following electrolytes from High to Low.
   (a) Hydrochloric acid (HCl)
   (b) Acetic acid (CH$_3$COOH)
   (c) Caustic soda (NaOH)
   (d) Potassium chloride (KCl)
   (e) Sodium chloride (NaCl)
   (f) Lithium chloride (LiCl)
2. Mark the correct statement(s).
   (a) Anions and cations migrate at the same rate
   (b) Anions always migrate faster than cations
   (c) Protons travel fastest in water
   (d) Cations always have the larger transport numbers than anions
3. A so-called isotonic saline solution consists of 9.00 g of common salt (NaCl, $M = 58.44$ g/mol) in 1.00 liter of aqueous solution. Accordingly, the solution has a weigh-in concentration of 154 mmol/L.
   Mark the correct specification(s)
   (a) The activity of the solution is equal to its weigh-in concentration
   (b) The activity coefficient of the solution is less than 1
   (c) The ionic strength of the solution corresponds to its weigh-in concentration
   (d) The activity of protons in the solution is lower than in pure water

11.14 Exercises

1. A lithium chloride solution ($LiCl \rightarrow Li^+ + Cl^-$) is electrolyzed. The anode and cathode are arranged as in a plate capacitor.
   The voltage between the anode and cathode is 60.8 V. The distance between the anode and cathode is 24.8 cm.
   (a) What is the field strength in the electrolyte?
   (b) Calculate the mobility u_+ of the cation.
   (c) Calculate the drift velocity v_+ of the cation
   (d) Calculate the transport number t_+ of the cation

   Calculate the specific conductivities κ_E and the pH values of the following acid solutions at 25 °C

(a) 1.00 mmol sulfuric acid in 1.00 L water (strong electrolyte)

$$\left(H_2SO_4 \rightarrow 2\,H^+ + SO_4^{2-} \,; K_{Kohlrausch} = 364 \, \frac{mS\ m^2}{mol\sqrt{\frac{mol}{L}}} \right)$$

(b) 1.00 mmol acetic acid in 1.00 liter water (weak electrolyte)

$$\left(HOAc \rightleftharpoons H^+ + OAc^- \,; K_a = 1.3 \cdot 10^{-5} \frac{mol}{L} \right)$$

Electrodes

12

12.1 Motivation

Batteries convert chemical energy into electrical energy. But where do we find voltage and current in a chemical reaction? To answer these questions, we need to look at electrodes. (The *motivational picture* of this chapter, Fig. 12.1, illustrates the electric potentials and the electromotive force in a Daniell cell).

12.2 What Happens When Electrons Are Transferred <u>into</u> the Electrolyte?

When we build a circuit with an electrolyte, charges have to be transferred between an electronic conductor and an ionic conductor.

We will discuss the phenomena in detail for the electrolysis of copper chloride solution (see Fig. 12.2).

At one electrode, electrons are transferred from the metal (electronic conductor) into the electrolyte. We call this electrode the cathode.

The transfer of electrons between the electronic conductor and the ionic conductor is called the electron transfer reaction.

$$\nu_{Ox}Ox + \nu_e e^- \underset{Ox}{\overset{Red}{\rightleftarrows}} \nu_{Red} Red \tag{12.1}$$

We always specify the electron transfer reaction with the oxidized species and the electrons on the left-hand side and the reduced species on the right-hand side.

Thus, in a cathodic current, the electron transfer reaction proceeds from left to right as a reduction.

J. S. Lauth, *Physical Chemistry in a Nutshell*,
https://doi.org/10.1007/978-3-662-67637-0_12

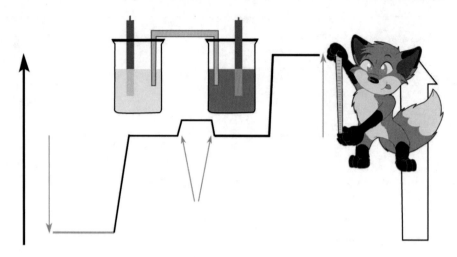

Fig. 12.1 How do we determine voltage and current in GALVANIC cells and electrolysis? (https://doi.org/10.5446/46042)

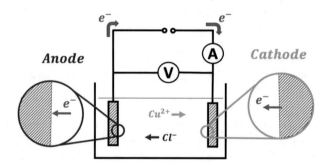

Fig. 12.2 Anodic and cathodic current during electrolysis of a copper chloride solution

$$\nu_{Ox}\,[Ox] + \nu_e\,e^- \quad \rightarrow \quad \nu_{Red}\,[Red]$$

$$Cu^{2+} + 2\,e^- \rightarrow Cu$$

Fig. 12.3 Copper deposition in copper chloride electrolysis as an example of a cathodic process

Cathode and reduction are therefore always coupled. If the cathodic reaction is forced, as in the electrolysis of copper chloride solution, the cathode is the negative terminal (see Fig. 12.3).

In a battery (GALVANIC cell) however, the cathodic reaction takes place naturally (unforced, "voluntarily"). Here, the cathode will be the positive terminal (see Table 12.1).

Table 12.1 Characterization of the cathode in a GALVANIC cell and in an electrolysis cell

Galvanic cell	Flow of electrons happens voluntarily	Reduction reaction predominates (e.g., metal deposition)	Flow of electrons from the metal into the electrolytic solution predominates	Positive terminal
Electrolysis cell	Flow of electrons is forced	Reduction reaction predominates (e.g., metal deposition)	Flow of electrons from the metal into the electrolytic solution predominates	Negative terminal

12.3   What Happens During the Transfer of Electrons <u>from</u> the Electrolyte?

At the other electrode in the circuit, electrons are transferred from the electrolyte into the metal. This is the anode and the electron transfer reaction proceeds here from right to left as oxidation.

In electrolysis cells, the anodic reaction is forced; in copper chloride electrolysis, this is the oxidation of chloride to chlorine. In electrolysis cells, the anode is therefore the positive terminal (see Fig. 12.4).

In GALVANIC cells, on the other hand, the anodic reaction takes place naturally (unforced, "voluntarily"); here the anode will be the negative terminal, such as the zinc electrode in a zinc–carbon battery (see Table 12.2).

12.4   How Much Conversion Takes Place at the Electrodes?

The laws of stoichiometry apply to the electron transfer reaction as to any other chemical reaction. Therefore, we may quantify conversion using the number of electrons involved.

$$\nu_{Ox} \, [Ox] + \nu_e \, e^- \overset{i_{cath}}{\underset{i_{an}}{\rightleftarrows}} \nu_{Red} \, [Red] \tag{12.6}$$

With this consideration, Michael FARADAY was able to specify the laws of electrolysis named after him: the amount of substance n converted at the electrodes does only depend on current I and on time t.

$$n = \frac{m}{M} = \frac{I \cdot t}{\nu_e \, F} \tag{12.7}$$

If we electrolyze for one day using a current of 1 A, 0.9 mol of electrons will flow.

$$\nu_{Ox}\,[Ox] + \nu_e\,e^- \quad \leftarrow \quad \nu_{Red}\,[Red]$$

$$Cl_2 + 2\,e^- \leftarrow 2\,Cl^-$$

Fig. 12.4 Chlorine formation in copper chloride electrolysis as an example of an anodic process

Table 12.2 Characterization of the anode in a GALVANIC cell and in an electrolysis cell

Galvanic cell	Flow of electrons happens voluntarily	Oxidation reaction predominates (e.g., metal dissolution)	Flow of electrons from the electrolytic solution into the metal predominates	Negative terminal
Electrolysis cell	Flow of electrons is forced	Oxidation reaction predominates (e.g., metal dissolution)	Flow of electrons from the electrolytic solution into the metal predominates	Positive terminal

$$\frac{I \cdot t}{F} = \frac{1\,A \cdot 86\,400\,s}{96\,485\,As/mol} = 0.9\,mol \tag{12.8}$$

These can produce, e.g., 0.45 mol copper

$$Cu^{2+} + 2\,e^- \rightarrow Cu \tag{12.9}$$

or 0.9 mol of silver at the cathode.

$$Ag^+ + e^- \rightarrow Ag \tag{12.10}$$

FARADAY'S law applies equally to anode and cathode; it applies to electrolytic and GALVANIC cells.

12.5 How Large Is the Potential Jump at the Metal/Electrolyte Phase Boundary?

The electron transfer reaction occurs at the metal/electrolyte phase boundary and should be formulated as a reversible reaction.

$$
\begin{array}{c}
\text{Red} \\
Cu^{2+} + 2\,e^- \rightleftarrows Cu \\
\text{Ox}
\end{array} \tag{12.11}
$$

Copper is a rather noble metal, therefore the equilibrium of a copper electrode lies on the right-hand side (see Fig. 12.5).

When we immerse copper metal in a copper chloride solution, the cathodic reaction initially predominates and the deposition of Cu^{2+} ions (cathodic current i_-) is faster than the dissolution of the copper (anodic current i_+). As a result, the copper metal becomes positively charged and the currents eventually become equal.

Fig. 12.5 Equilibrium
potential of a copper electrode

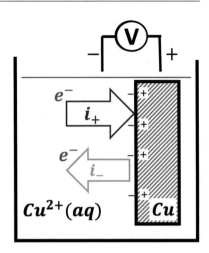

In equilibrium, the copper is positively charged and the electrolyte is negatively charged. Anodic and cathodic current densities are equal and correspond to the so-called exchange current density i_0.

$$|i_-| = i_+ = i_0 \tag{12.12}$$

We call the potential difference in equilibrium between metal and electrolyte redox potential $E_{red/ox}$.

$$E_{red/ox} = \varphi_{Me} - \varphi_{El} \tag{12.13}$$

$E_{red/ox}$ is characteristic for each electrode; for the copper electrode in the standard state (1 mol/L; pure copper) it will be

$$E^{\circ}{}_{Cu/Cu^{2+}} = +0.34 \text{ V} \tag{12.14}$$

If we repeat the experiment with zinc metal and zinc chloride solution, we obtain a negative redox potential:

$$E^{\circ}{}_{Zn/Zn^{2+}} = -0.74 \text{ V} \tag{12.15}$$

Zinc being a non-noble metal, the equilibrium of the electron transfer reaction is on the left-hand side.

12.6   How Does Redox Potential Depend on Concentration?

The potential of an electrode depends not only on which electron transfer reaction takes place, but also on which concentrations are present in the individual case (see Fig. 12.6).

Fig. 12.6 Concentration dependence of the redox potential of a copper electrode

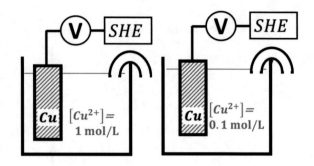

In a 1-molar solution, the redox potential of copper is 0.34 V; in a 0.1-molar solution, it is only 0.31 V (all values measured against the standard hydrogen electrode SHE).

The concentration dependence of the redox potential was specified by Walter NERNST in the famous equation named after him.

$$E_{Red/Ox} = E°_{Red/Ox} + \frac{RT}{\nu_e F} \ln \frac{[Ox]^{\nu_{Ox}}}{[Red]^{\nu_{Red}}} \qquad (12.16)$$

According to NERNST redox potential, $E_{Red/Ox}$ consists of the standard potential $E°_{Red/Ox}$ and a concentration-dependent term (incl. NERNST factor $\frac{RT}{\nu_e F}$).

The standard potentials $E°_{Red/Ox}$ are tabulated in the electrochemical series.

When calculating redox potentials, we must keep in mind that all parameters that appear in the electron transfer reaction also will appear in the NERNST equation:

$$\begin{array}{c} \text{Red} \\ \nu_{Ox}Ox + \nu_e e^- \ \rightleftarrows \ \nu_{Red} \ \text{Red} \\ \text{Ox} \end{array} \qquad (12.17)$$

The number of electrons involved ν_e can be found in the NERNST factor

$$\frac{RT}{\nu_e F} \qquad (12.18)$$

The concentrations of all oxidized species can be found in the numerator and the concentration of all reduced species can be found in the denominator in the argument of the logarithm.

12.7   How Do We Use the Electrochemical Series? (see Table 12.3)

The electrochemical series is a listing of redox equilibria ordered by their standard redox potentials $E°_{Red/Ox}$. Listed at the top are the redox equilibria with positive redox potential, i.e. high electron affinity (great "electron hunger").

Table 12.3 Section of the electrochemical series

Oxid. form/red. form	Durchtrittsreaktion	E^0_{redox} in V
Cl_2/Cl^-	$Cl_2(g) + 2\ e^- \rightleftharpoons 2\ Cl^-(aq)$	+1.36
$O_2, H^+/H_2O$	$O_2(g) + 4\ H^+(aq) + 4\ e^- \rightleftharpoons 2\ H_2O(l)$	+1.23
Ag^+/Ag	$Ag^+(aq) + e^- \rightleftharpoons Ag(s)$	+0.80
Cu^{2+}/Cu	$Cu^{2+}(aq) + 2\ e^- \rightleftharpoons Cu(s)$	+0.34
$AgCl/Ag, Cl^-$	$AgCl(s) + e_- \rightleftharpoons Ag(s) + Cl^-(aq)$	+0.22
H^+/H_2	$2\ H^+(aq) + 2\ e^- \rightleftharpoons H_2(g)$	0.00
Fe^{2+}/Fe	$Fe^{2+}(aq) + 2\ e^- \rightleftharpoons Fe(s)$	−0.44
Zn^{2+}/Zn	$Zn^{2+}(aq) + 2\ e^- \rightleftharpoons Zn(s)$	−0.76
Mg^{2+}/Mg	$Mg^{2+}(aq) + 2\ e^- \rightleftharpoons Mg(s)$	−2.36

At the bottom of the list, we find the redox equilibria with low redox potential, i.e. those that give up electrons very easily (high "electron pressure").

Thus, the redox equilibria at the top will be strongly oxidizing; the redox equilibria at the bottom are strongly reducing. If we combine two half-reactions from the electrochemical series, the electrons will always flow naturally only from the bottom to the top.

12.8 How Do We Describe a First Type Electrode (Metal/Metal Salt)?

Let us specify the NERNST equation for a copper electrode

$$Cu\ (s)/Cu^{2+}(aq) \tag{12.19}$$

We first specify the electron transfer reaction.

$$\overset{\text{Red}}{Cu^{2+}\ (aq) + 2\ e^- \underset{\text{Ox}}{\rightleftarrows} Cu(s)} \tag{12.20}$$

In the electrochemical series, we find the corresponding standard potential $E°_{Red/Ox}$. In the NERNST factor two electrons have to be taken into account.

In the numerator of the argument of the logarithm, we plug in the concentration of the oxidized species, i.e. the Cu^{2+} ions. The denominator depicts the concentration of the reduced species, i.e. the copper metal.

$$E_{Cu/Cu^{2+}} = E°_{Cu/Cu^{2+}} + \frac{RT}{2\,F} \ln \frac{[Cu^{2+}]}{[Cu]} \tag{12.21}$$

The concentration conventions of thermodynamics apply here (just as, for example, with the thermodynamic equilibrium constant): solids and liquids are quantified using their mole fraction, for gaseous substances we have to use partial pressure in bar, and dissolved substances are quantified using molarity.

$$[Cu^{2+}] = \frac{c_{Cu^{2+}}}{mol/L} \qquad (12.22)$$

$$[O_2] = \frac{p_{O_2}}{bar} \qquad (12.23)$$

$$[Cu] = \frac{x_{Cu}}{mol/mol} \qquad (12.24)$$

(for exact calculations, the corresponding activities must be used)

The dependence of potential on the concentration is logarithmic. The NERNST factor combined with a logarithm argument of 10 gives the value

$$\frac{RT}{F} \ln(10) \approx 59 \text{ mV} \qquad (12.25)$$

This means that for the copper electrode with a tenfold increase of the copper ion concentration the potential changes by $\frac{59 \text{ mV}}{2} \approx 30$ mV.

A copper electrode is a so-called electrode of the first kind—the electrode material itself is the reduced species; only one phase boundary occurs.

12.9   How Do We Describe a Gas Electrode?

A gas electrode has a somewhat more complicated structure than an electrode of the first kind. We are dealing with three phases in total here (electrolyte, gas, metal); the metal is often an inert metal such as platinum or graphite (see Fig. 12.7).

We can find oxygen electrodes in zinc–air batteries or in fuel cells.

$$\overset{\text{Red}}{4\,H^+\,(aq) + O_2(g) + 4\,e^- \;\rightleftarrows\; 2H_2O(l)} \qquad (12.26)$$
$$\underset{\text{Ox}}{}$$

Fig. 12.7 Oxygen electrode as an example of a gas electrode

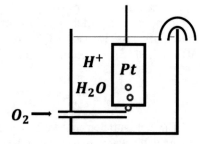

We look up the electron transfer reaction in the electrochemical series; the standard potential of 1.23 V tells us that we are dealing with a very strongly oxidized redox equilibrium.

$$E^{\circ}{}_{H_2O/O_2} = +1.23 \text{ V} \tag{12.27}$$

In the NERNST equation, we consider four electrons exchanged in the NERNST *factor*. In the argument of the logarithm, we plug in the proton concentration to the fourth power and the oxygen concentration into the numerator; in the denominator, we have to fill in the water concentration squared.

$$E_{O_2/H^+} = E^{\circ}{}_{O_2/H^+} + \frac{RT}{4F} \ln \frac{[H^+]^4[O_2]}{[H_2O]^2} \tag{12.28}$$

The proton concentration must be given in mol/L; the oxygen concentration in bar. For the concentration of water, we would have to use the mole fraction, but this can be set equal to 1 as a good approximation. The potential of the oxygen electrode is strongly pH dependent. A pH change of 1 results in a potential change of 59 mV.

12.10   How Do We Determine Sign and Magnitude of the Open-Circuit Voltage ("EMF")?

We can combine any two electrodes to form a battery; we then speak of GALVANIC cells, electric cells, or GALVANIC elements. Classically, the DANIELL cell consists of a copper electrode and a zinc electrode (see Fig. 12.8).

To calculate the voltage supplied by this battery (more precisely: the open-circuit voltage or *emf*), we need to formulate the electron transfer reactions of both electrodes and determine their redox potentials using NERNST's equation.

$$Zn^{2+} + 2e^- \leftarrow_{Ox} Zn \tag{12.29}$$

Fig. 12.8 DANIELL cell (GALVANIC cell consisting of a copper electrode and zinc electrode for the reversible reduction of copper ions with zinc)

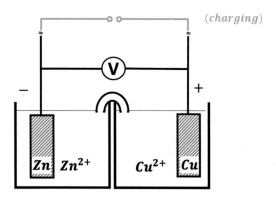

$$Cu^{2+} + 2e^{-} \xrightarrow{\text{Red}} Cu \tag{12.30}$$

The electrode with the smaller redox potential is the anode, this is where oxidation occurs. The electrode with the higher redox potential is the cathode, this is where the reduction takes place. The open-circuit voltage of the GALVANIC cell corresponds to the difference of the redox potentials

$$\Delta E_{\text{Galv}} = E_{\text{Cath}} - E_{\text{An}} \tag{12.31}$$

For the classical DANIELL cell consisting of the two electrodes in the standard state, we obtain an open-circuit voltage of

$$\Delta E_{\text{Galv}} = (+0.34 \text{ V}) - (-0.76 \text{ V}) = 1.10 \text{ V} \tag{12.32}$$

12.11 How Is a Spontaneous Redox Reaction Different from a GALVANIC Cell?

In each battery a redox reaction takes place, which can be obtained by a simple combination of the electron transfer reactions. In the DANIELL cell, this is simply the reduction of copper ions by zinc:

$$Cu^{2+}(aq) + Zn(s) \rightarrow Zn^{2+}(aq) + Cu(s) \tag{12.33}$$

If we combine zinc and copper ions without a GALVANIC cell, this redox reaction will also take place ("spontaneous redox reaction"), but then we cannot obtain electricity from the reaction.

The heat released during the spontaneous redox reaction corresponds to the enthalpy of reaction $\Delta_R H$

$$q_{p,\text{spon.}} = \Delta_R H \tag{12.34}$$

In the DANIELL process, these are (as we can calculate, e.g., from thermodynamic tables)

$$\Delta_R H = -217 \frac{\text{kJ}}{\text{mol}} \tag{12.35}$$

[Useful] work, as with any spontaneous reaction, is not implemented.

$$w_{p,\text{spon.}} = 0 \tag{12.36}$$

If the same reaction takes place in a DANIELL cell, oxidation and reduction are spatially separated. Electrons must flow and electric work can be gained, ideally we will get the maximum—the reversible—electric work, which corresponds to the free enthalpy

$$w_{el.,rev.} = \Delta_R G \tag{12.37}$$

In the *DANIELL process*, these are (as we can calculate from thermodynamic tables)

$$\Delta_R G = -212 \frac{kJ}{mol} \tag{12.38}$$

The reversible heat of reaction can be calculated using reaction entropy.

$$q_{p,rev.} = T \cdot \Delta_R S \tag{12.39}$$

The reversible reaction is only weakly exothermic (in contrast to the spontaneous redox reaction).

$$T \, \Delta_R S = -6 \frac{kJ}{mol} \tag{12.40}$$

We call the quotient of free enthalpy and enthalpy the efficiency of the *GALVANIC* cell.

$$\eta = \frac{\Delta_R G}{\Delta_R H} \tag{12.41}$$

$$\eta = \frac{-212}{-217} = 0.98 \tag{12.42}$$

Whereas the entire enthalpy is released as heat in the spontaneous reaction, we can gain a very large fraction of this enthalpy as work performing the reaction in reversible mode.

This is the great advantage of using reversible chemical reactions (batteries) compared to spontaneous chemical reactions (combustion) to obtain useful work.

The heat obtained in spontaneous processes can only be converted into work with a relatively low efficiency (approx. 30%; see CARNOT process).

In contrast, the efficiency of *GALVANIC* cell is in typically 90% or higher.

12.12   How Large Is the Potential Jump at a Semipermeable Membrane?

The combination of metal and electrolyte thus leads to a redox potential. The combination of two electrolytes can also lead to a potential.

Figure 12.9 shows two solutions with different pH values separated by a membrane. The membrane is permeable to protons. Protons now diffuse from the more concentrated solution through the membrane into the more dilute solution and will charge this side.

Theory provides an equation for the membrane potential similar to the *NERNST* equation.

Fig. 12.9 Potential at a
semipermeable membrane
between two solutions with
different pH values

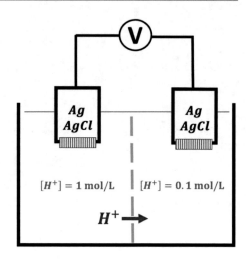

$$\Delta_{\text{Mem}}\varphi = \varphi(II) - \varphi(I) = -\frac{RT}{z_i F} \ln \frac{[i]^{II}}{[i]^{I}} \tag{12.43}$$

Membrane potentials are basic for many electrochemical measurements, for example, measurement of pH values.

Furthermore, membrane potentials play an important role in biology with the conduction of nerve impulses.

12.13 Summary

The combination of an electronic conductor with an ion conductor results in an electrode. An electrode is described by the electron transfer reaction

$$\overset{i_{\text{cath}}}{\underset{i_{\text{an}}}{\nu_{\text{Ox}} \left[\text{Ox}\right] + \nu_e\, e^- \; \rightleftarrows \; \nu_{\text{Red}}\left[\text{Red}\right]}} \tag{12.44}$$

and by the redox potential.

The redox potential does depend on concentration. This is specified by the NERNST equation.

$$E_{\text{red/ox}} = E^{\circ}{}_{\text{red/ox}} + \frac{R\,T}{\nu_e F}\, \ln \frac{\left[\text{Ox}\right]^{\nu_{\text{Ox}}}}{\left[\text{Red}\right]^{\nu_{\text{Red}}}} \tag{12.45}$$

The amount of substance converted at an electrode is specified by FARADAY'S law of electrolysis.

$$n = \frac{m}{M} = \frac{I \cdot t}{\nu_e F} \tag{12.46}$$

We can combine electrodes to form GALVANIC cells. The open-circuit voltage or emf of these GALVANIC cells corresponds to the difference of the redox potentials.

$$\Delta E_{\text{Galv}} = E_{\text{Cath}} - E_{\text{An}} \tag{12.47}$$

12.14 Test Questions

QUIZ

1. Mark the correct statement(s).
   (a) The anode is always the negative terminal
   (b) The reduction takes place at the cathode
   (c) The cathode always has the more positive potential
   (d) In electrolysis, the cathode is the positive terminal
   (e) In a GALVANIC cell, the anode is the negative terminal

Fig. 12.10 Three electrolytic cells in series

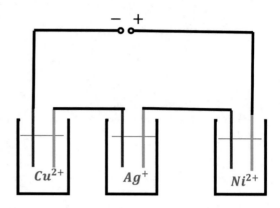

2. Determine the open-circuit voltage of a *GALVANIC* cell consisting of an iron electrode $\left(E°\frac{Fe}{Fe^{2+}} = -0.44 \text{ V}\right)$ and a silver electrode $\left(E°_{Ag/Ag^+} = 0.80 \text{ V}\right)$ at standard conditions
   (a) 1.24 V
   (b) 0.36 V
   (c) 0.62 V

3. Three solutions (copper sulfate, silver nitrate, nickel chloride) are electrolyzed simultaneously in a series circuit. Mark the correct statement(s) (Fig. 12.10).

   (a) The current in all three solutions is identical
   (b) The voltage at all three solutions is identical
   (c) The deposited amount of silver is twice as high as the deposited amount of copper

12.15 Exercises

1. In a zinc–carbon battery, the anode consists of 2.81 g of zinc. The voltage supplied by the battery is 0.99 V on average.

   (a) What is the maximum amount of charge that the battery can supply?
   (b) What is the maximum electrical energy that the battery can supply?
2. 1 mole of hydrogen is converted to liquid water at 298 K in an ideally operating
   fuel cell (PEMFC).
   (a) Calculate the (ideal) efficiency ($\eta = \Delta G / \Delta H$) of this fuel cell.
   (b) How much electrical energy w_{el} and how much heat q are emitted?

3. A semipermeable cell membrane (permeable to potassium ions, impermeable to
   all other ions, i.e. $t_+ = 1$) separates two solutions of potassium concentrations
   155 mmol/L (cell interior) and 4 mmol/L (cell exterior). Calculate the membrane
   potential $\Delta_{Mem}\varphi$ between the solutions at 37 °C.

4. An oxygen electrode (pH $= 7$, $p(O_2) = 100$ kPa) and a silver electrode (c
   $(Ag^+) = 1.00\,10^{-4}$ mol/L) are connected at 298 K.
   (a) Calculate the open-circuit voltage ΔE of this GALVANIC cell
   (b) Which electrode is the anode; which electrode is the positive terminal of
   the cell?

Service Section (Appendix)

13.1 Solutions of the Tests and Exercises

The motivational picture of this chapter, Fig. 13.1, illustrates the submission of an assignment

Fig. 13.1 Solutions to the test questions and exercises

J. S. Lauth, *Physical Chemistry in a Nutshell*, https://doi.org/10.1007/978-3-662-67637-0_13

Chapter 1 "Changes of State"

Solutions to the Test Questions

1. 0 Degrees of freedom
2. $w > 0; q < 0$
3. $w < 0; q > 0$
4. b and d
5. $w > 0; q < 0$

Solutions to the Exercises

1.

$$q_{\text{sensible,Ice}} = c_p(s) \cdot m \cdot \left(T_f - T_i\right) = 2.03 \frac{\text{kJ}}{\text{kg K}} 0.018 \text{ kg } (25 \text{ K}) = 0.914 \text{ kJ}$$

$$q_{\text{sensible,Water}} = c_p(l) \cdot m \cdot (T_f - T_i) = 4.18\,\frac{\text{kJ}}{\text{kg K}}\,0.018\ \text{kg}\ (100\ \text{K}) = 7.52\ \text{kJ}$$

$$q_{\text{sensible,water vapor}} = 1.84\,\frac{\text{kJ}}{\text{kg K}}\,0.018\ \text{kg}\ (25\ \text{K}) = 0.828\ \text{kJ}$$

$$q_{\text{latent,melting}} = \Delta_{\text{fus}}H \cdot n = 6.01\,\frac{\text{kJ}}{\text{mol}} \cdot 1.00\ \text{mol} = 6.01\ \text{kJ}$$

$$q_{\text{latent,vaporizing}} = \Delta_{\text{vap}}H \cdot n = 40.67\,\frac{\text{kJ}}{\text{mol}} \cdot 1.00\ \text{mol} = 40.7\ \text{kJ}$$

$$q_{\text{total}} = \sum q_{\text{sensible}} + q_{\text{latent}} = 56.0\ \text{kJ}$$

2.

$$q_{\text{Polystyrene}} = c_p(P) \cdot m(P) \cdot \left(T_{\text{eq}} - T_i(P)\right) = 2.3\,\frac{\text{kJ}}{^\circ\text{C}}\left(T_{\text{eq}} - 50.0\,^\circ\text{C}\right)$$

$$q_{\text{Water}} = c_p(W) \cdot m(W) \cdot \left(T_{\text{eq}} - T_i(W)\right) = 41.8\,\frac{\text{kJ}}{^\circ\text{C}}\left(T_{\text{eq}} - 20.0\,^\circ\text{C}\right)$$

$$q_{\text{Water}} = -q_{\text{Polystyrene}} \quad (\text{Basic equation of calorimetry})$$

$$2.3\,\frac{\text{kJ}}{^\circ\text{C}}\left(T_{\text{eq}} - 50.0\,^\circ\text{C}\right) = -41.8\,\frac{\text{kJ}}{^\circ\text{C}}\left(T_{\text{eq}} - 20.0\,^\circ\text{C}\right)$$

$$T_{\text{eq}} = \frac{115\ \text{kJ} + 836\ \text{kJ}}{41.8\,\frac{\text{kJ}}{^\circ\text{C}} + 2.3\,\frac{\text{kJ}}{^\circ\text{C}}} = 21.56\,^\circ\text{C}$$

$$q_{\text{Water}} = 41.8\,\frac{\text{kJ}}{^\circ\text{C}}\left(21.56 - 20.0\,^\circ\text{C}\right) = 65.4\ \text{kJ}$$

3. Solution looks similar to the isotherm in Fig. 2.12

Chapter 2 "Gases"

Solutions to the Test Questions

1. 0.5 liters
2. Density $= 1.3$ g/L; Volume $= 22.4$ liters
3. Ar and O_2 have the same average energy; Ar is slowest (on average)
4. Mean velocity ~ 1375 km/h (854 mph); mean free path ~ 0.075 μm
5. The surface tension is zero; the heat of vaporization is zero
6. Left curve (red) corresponds to nitrogen (N_2); right curve (green) corresponds to helium (He)

Solutions to the Exercises

1.

$$n = \frac{pV}{RT} = \frac{0.500 \cdot 2340 \text{ Pa} \cdot 1000 \text{ m}^3}{8.314 \frac{\text{Pa m}^3}{\text{K mol}} \cdot 293 \text{ K}} = 480 \text{ mol}$$

$$m = n\,M = 480 \text{ mol} \cdot 18 \frac{\text{g}}{\text{mol}} = 8.65 \text{ kg}$$

2.

$$n = \frac{pV}{RT} = \frac{123 \text{ kPa} \cdot 2.00 \text{ L}}{8.314 \frac{\text{kPa L}}{\text{K mol}} \cdot 291 \text{ K}} = 101.7 \text{ mmol}$$

$$\overline{V} = \frac{V}{n} = \frac{RT}{p} = \frac{8.314 \frac{\text{kPa L}}{\text{K mol}} \cdot 291 \text{ K}}{123 \text{ kPa}} = 19.7 \frac{\text{L}}{\text{mol}}$$

$$M = \frac{m}{n} = \frac{6.00 \text{ g}}{0.1017 \text{ mol}} = 59.0 \frac{\text{g}}{\text{mol}}$$

Chapter 3 "Physical Equilibria"

Solutions to the Test Questions

1. a, b, and c
2. a, c, and d
3. a: at the point of greatest slope
   b: at the point of greatest curvature

Solutions to the Exercises

1.

$$\frac{dq}{A\,dt} = -A\,\lambda\,\frac{dT}{dx} \tag{13.1}$$

$$\frac{dq}{dt} = -1.00 \text{ m}^2 \cdot 0.76 \frac{\text{J}}{\text{°C m s}} \frac{18.0\,\text{°C} - 20.0\,\text{°C}}{0.00400 \text{ m}} = 380 \frac{\text{J}}{\text{s}} = 0.38 \text{ kW}$$

2.

$$\eta_{\text{Carnot}} = \frac{T_{\text{high}} - T_{\text{low}}}{T_{\text{high}}} = \frac{298 \text{ K} - 273 \text{ K}}{298 \text{ K}} = 0.0839 \ (8.39\%)$$

$$\eta_{\text{Carnot}} = -\frac{w_{\text{rev}}}{q_{\text{high}}}$$

$$w_{\text{rev}} = -\eta_{\text{Carnot}} \cdot q_{\text{high}} = -0.0839 \cdot (-500 \text{ kJ}) = 41.9 \text{ kJ}$$

3.

$$\eta_{\text{Carnot}} = \frac{T_{\text{high}} - T_{\text{low}}}{T_{\text{high}}} = \frac{773 \text{ K} - 373 \text{ K}}{773 \text{ K}} = 0.517 \ (51.7\%)$$

$$\eta = 0.80 \cdot 0.517 = 0.414 \ (41.4\%)$$

$$\eta = -\frac{w}{q_{\text{high}}}$$

$$q_{high} = -\frac{w}{\eta} = -\frac{(-50.0 \text{ MJ})}{0.414} = 121 \text{ MJ}$$

$$q_{\text{high}} + q_{\text{low}} + w = 0$$

$$q_{\text{low}} = -q_{\text{high}} - w = -(121 \text{ MJ}) - (-50.0 \text{ MJ}) = -71 \text{ MJ}$$

Chapter 4 "Affinity"

Solutions to the Test Questions

1. b and e
2. b, e, and f
3. a: enthalpy increases, entropy increases
   b: enthalpy increases, entropy increases
       c: enthalpy remains the same, entropy increases
       d: enthalpy decreases, entropy decreases

Solutions to the Exercises

1.

$$\Delta_r H^\circ = \Delta_f H^\circ \,(\text{Products}) - \Delta_f H^\circ \,(\text{Reactants})$$

$$\Delta_r H^{\circ} = \Delta_f H^{\circ} (\text{Ca(OH)}_2) - (\Delta_f H^{\circ} (\text{CaO}) + \Delta_f H^{\circ} (\text{H}_2\text{O(l)}))$$

$$\Delta_r H^{\circ} = \left(- 986 \frac{\text{kJ}}{\text{mol}}\right) - \left(\left(- 635.1 \frac{\text{kJ}}{\text{mol}}\right) + \left(- 285.84 \frac{\text{kJ}}{\text{mol}}\right)\right) = - 66 \frac{\text{kJ}}{\text{mol}}$$

2.

$$\Delta_r H^{\circ} = \Delta_f H^{\circ} (\text{Products}) - \Delta_f H^{\circ} (\text{Reactants})$$

$$\Delta_r H^{\circ} = \left(4 \cdot \Delta_f H^{\circ} (\text{H}_2\text{O(g)}) + 2 \cdot \Delta_f H^{\circ} (\text{N}_2) + \Delta_f H^{\circ} (\text{O}_2)\right) \\ - \left(2 \cdot \Delta_f H^{\circ} (\text{NH}_4\text{NO}_3)\right)$$

$$\Delta_r H^{\circ} = \left(4 \cdot \left(- 241.83 \frac{\text{kJ}}{\text{mol}}\right) + 2 \cdot 0 + 0\right) - \left(2 \cdot \left(- 365.6 \frac{\text{kJ}}{\text{mol}}\right)\right) = $$

$$- 236.1 \frac{\text{kJ}}{\text{mol}}$$

$$\Delta_r S^{\circ} = \Delta_f S^{\circ} (\text{Products}) - \Delta_f S^{\circ} (\text{Reactants})$$

$$\Delta_r H^{\circ} = (4 \cdot S^{\circ} (\text{H}_2\text{O(g)}) + 2 \cdot S^{\circ} (\text{N}_2) + S^{\circ} (\text{O}_2)) - (2 \cdot S^{\circ} (\text{NH}_4\text{NO}_3))$$

$$\Delta_r H^{\circ} = \left(4 \cdot \left(188.72 \frac{\text{J}}{\text{mol}}\right) + 2 \cdot 191.5 \frac{\text{J}}{\text{mol K}} + 205 \frac{\text{J}}{\text{mol K}}\right) \\ - \left(2 \cdot \left(- 151 \frac{\text{J}}{\text{mol K}}\right)\right) = 1040.88 \frac{\text{J}}{\text{mol K}}$$

$$\Delta_r G^{\circ} = \Delta_r H^{\circ} - T \Delta_r S^{\circ}$$

$$\Delta_r G^{\circ} = \left(- 236.1 \frac{\text{kJ}}{\text{mol}}\right) - 371.95 \text{ K} \left(+1040.88 \frac{\text{J}}{\text{mol K}}\right) = - 623.28 \frac{\text{kJ}}{\text{mol}}$$

$$n = \frac{m}{M} = \frac{1510 \text{ g}}{80.04 \frac{\text{g}}{\text{mol}}} = 18.87 \text{ mol NH}_4\text{NO}_3 = 9.43 \text{ mol} * \text{reaction equation}$$

$$q = n \cdot \Delta_r H^{\circ} = 9.43 \text{ mol} \cdot \left(- 236.1 \frac{\text{kJ}}{\text{mol}}\right) = - 2227 \text{ kJ} = - 2.227 \text{ MJ}$$

3.

$$2 \text{ H}_3\text{C} - \text{CH}_3 + 7 \text{ O} = \text{O} \rightarrow 4 \text{ O} = \text{C} = \text{O} + 6 \text{ H} - \text{O} - \text{H}$$

$$\Delta_r H^\circ \approx \sum \langle H_{bond} \rangle (\text{Products}) - \sum \langle H_{bond} \rangle (\text{Reactants})$$

$$\Delta_r H^\circ \approx (8 \cdot \langle H_{bond} \rangle (C=O) + 12 \cdot \langle H_{bond} \rangle (O-H))$$
$$- (2 \cdot \langle H_{bond} \rangle (C-C) + 12 \cdot \langle H_{bond} \rangle (C-H) + 7 \cdot \langle H_{bond} \rangle (O=O))$$

$$\Delta_r H^\circ \approx \left(8 \cdot \left(-799 \frac{kJ}{mol} \right) + 12 \cdot \left(-463 \frac{kJ}{mol} \right) \right)$$
$$- \left(2 \cdot \left(-346 \frac{kJ}{mol} \right) + 12 \cdot \left(-413 \frac{kJ}{mol} \right) + 7 \cdot \left(-498 \frac{kJ}{mol} \right) \right) = -2814 \frac{kJ}{mol}$$

$$n = \frac{m}{M} = \frac{1000\ g}{30.07 \frac{g}{mol}} = 33.26\ \text{mol}\ C_2H_6 = 16.63\ \text{mol} * \text{reaction equation}$$

$$q = n \cdot \Delta_r H^\circ \approx 16.63\ \text{mol} \cdot \left(-2814 \frac{kJ}{mol} \right) \approx -47\ MJ$$

Chapter 5 "Chemical Equilibria"

Solutions to the Test Questions

1. a and b
2. a: high temperature, low pressure
   b: low temperature, high pressure
        c: low temperature, high pressure
3. c
4. c

Solutions to the Exercises

1. a

$$\Delta_r H^\circ = \Delta_f H^\circ \, (\text{Products}) - \Delta_f H^\circ \, (\text{Reactants})$$

$$\Delta_r H^\circ = \left(\Delta_f H^\circ \, (\text{CO}_2) + \Delta_f H^\circ \, (\text{H}_2)\right) - \left(\Delta_f H^\circ \, (\text{CO}) + \Delta_f H^\circ \, (\text{H}_2\text{O(g)})\right)$$

$$\Delta_r H^\circ = \left(\left(-393.77\,\frac{\text{kJ}}{\text{mol}}\right) + 0\,\frac{\text{kJ}}{\text{mol}}\right) - \left(\left(-137.2\,\frac{\text{kJ}}{\text{mol}}\right) + \left(-241.83\,\frac{\text{kJ}}{\text{mol}}\right)\right)$$

$$= -41.32\,\frac{\text{kJ}}{\text{mol}}$$

b

$$\Delta_r G^\circ = \Delta_f G^\circ \, (\text{Products}) - \Delta_f G^\circ \, (\text{Reactants})$$

$$\Delta_r G^\circ = \left(\Delta_f G^\circ \, (\text{CO}_2) + \Delta_f G^\circ \, (\text{H}_2)\right) - \left(\Delta_f G^\circ \, (\text{CO}) + \Delta_f G^\circ \, (\text{H}_2\text{O(g)})\right)$$

$$\Delta_r G^\circ = \left(\left(-394.4\,\frac{\text{kJ}}{\text{mol}}\right) + 0\,\frac{\text{kJ}}{\text{mol}}\right) - \left(\left(-110.62\,\frac{\text{kJ}}{\text{mol}}\right) + \left(-228.6\,\frac{\text{kJ}}{\text{mol}}\right)\right)$$

$$= -28.6\,\frac{\text{kJ}}{\text{mol}}$$

2.

$$\Delta_r H^\circ = \Delta_f H^\circ \,(\text{Products}) - \Delta_f H^\circ \,(\text{Reactants})$$

$$\Delta_r H^\circ = \left(\Delta_f H^\circ \,(\text{CO}_2) + \Delta_f H^\circ \,(\text{CaO})\right) - \Delta_f H^\circ \,(\text{CaCO}_3)$$

$$\Delta_r H^\circ = \left(\left(-393.77\frac{\text{kJ}}{\text{mol}}\right) + \left(-635.1\frac{\text{kJ}}{\text{mol}}\right)\right) - \left(-1212.0\frac{\text{kJ}}{\text{mol}}\right) =$$

$$+\,183.1\frac{\text{kJ}}{\text{mol}}$$

$$\Delta_r S^\circ = S^\circ \,(\text{Products}) - S^\circ \,(\text{Reactants})$$

$$\Delta_r S^\circ = (S^\circ \,(\text{CO}_2) + S^\circ \,(\text{CaO})) - S^\circ \,(\text{CaCO}_3)$$

$$\Delta_r S^\circ = \left(\left(213.86\frac{\text{J}}{\text{mol K}}\right) + \left(39.7\frac{\text{J}}{\text{mol K}}\right)\right) - \left(92.9\frac{\text{J}}{\text{mol K}}\right) = +\,160.7\frac{\text{J}}{\text{mol K}}$$

$$T_{\text{floor}} = \frac{\Delta_{rxn} H}{\Delta_{rxn} S} = \frac{183100\frac{\text{J}}{\text{mol}}}{160.7\frac{\text{J}}{\text{mol K}}} = 1140 \text{ K } (867\,^\circ\text{C})$$

3.

$$\Delta_r G^\circ = \Delta_f G^\circ \,(\text{Products}) - \Delta_f G^\circ \,(\text{Reactants})$$

$$\Delta_r G^\circ = \Delta_f G^\circ \,(\text{H}_2\text{O}) - \left(\Delta_f G^\circ \,(\text{H}^+) + \Delta_f G^\circ \,(\text{OH}^-)\right)$$

$$\text{or } \Delta_r G^\circ = \mu^\circ \,(\text{H}_2\text{O}) - (\mu^\circ \,(\text{H}^+) + \mu^\circ \,(\text{OH}^-))$$

$$\Delta_r G^\circ = \left(-237.1\frac{\text{kJ}}{\text{mol}}\right) - \left(\left(0\frac{\text{kJ}}{\text{mol}}\right) + \left(-157.2\frac{\text{kJ}}{\text{mol}}\right)\right) = -\,79.9\frac{\text{kJ}}{\text{mol}}$$

or:

$$\Delta_r H^\circ = \Delta_f H^\circ \,(\text{Products}) - \Delta_f H^\circ \,(\text{Reactants})$$

$$\Delta_r H^\circ = \Delta_f H^\circ \,(\text{H}_2\text{O}) - \left(\Delta_f H^\circ \,(\text{H}^+) + \Delta_f H^\circ \,(\text{OH}^-)\right)$$

$$\Delta_r H^\circ = \left(-285.84\frac{\text{kJ}}{\text{mol}}\right) - \left(\left(0\frac{\text{kJ}}{\text{mol}}\right) + \left(-230\frac{\text{kJ}}{\text{mol}}\right)\right) = -\,56\frac{\text{kJ}}{\text{mol}}$$

$$\Delta_r S^\circ = S^\circ \,(\text{Products}) - S^\circ \,(\text{Reactants})$$

$$\Delta_r S^\circ = S^\circ \,(\text{H}_2\text{O}) - (S^\circ \,(\text{H}^+) + S^\circ \,(\text{OH}^-))$$

$$\Delta_r S^\circ = \left(69.9 \, \frac{\text{J}}{\text{mol K}}\right) - \left(\left(0 \, \frac{\text{J}}{\text{mol K}}\right) + \left(-10.8 \, \frac{\text{J}}{\text{mol K}}\right)\right) = +80.7 \, \frac{\text{J}}{\text{mol K}}$$

$$\Delta_r G^\circ = \Delta_r H^\circ - T\Delta_r S^\circ$$

$$\Delta_r G^\circ = \left(-57 \, \frac{\text{kJ}}{\text{mol}}\right) - 298.15 \, \text{K} \left(+80.7 \, \frac{\text{J}}{\text{mol K}}\right) = -80 \, \frac{\text{kJ}}{\text{mol}}$$

4.

$$\Delta_r G^\circ = \Delta_f G^\circ \,(\text{Products}) - \Delta_f G^\circ \,(\text{Reactants})$$

$$\Delta_r G^\circ = 2 \cdot \Delta_f G^\circ \,(\text{CO}) - \left(\Delta_f G^\circ \,(\text{CO}_2) + \Delta_f G^\circ \,(\text{C(s, graphite)})\right)$$

$$\text{or } \Delta_r G^\circ = 2 \cdot \mu^\circ \,(\text{CO}) - (\mu^\circ \,(\text{CO}_2) + \mu^\circ \,(\text{C(s, graphite)}))$$

$$\Delta_r G^\circ = 2 \cdot \left(-137.2 \, \frac{\text{kJ}}{\text{mol}}\right) - \left(\left(-394.4 \, \frac{\text{kJ}}{\text{mol}}\right) + \left(0 \, \frac{\text{kJ}}{\text{mol}}\right)\right) = +120 \, \frac{\text{kJ}}{\text{mol}}$$

at 25 °C
or:

$$\Delta_r H^\circ = \Delta_f H^\circ \,(\text{Products}) - \Delta_f H^\circ \,(\text{Reactants})$$

$$\Delta_r H^\circ = 2 \cdot \Delta_f H^\circ \,(\text{CO}) - \left(\Delta_f H^\circ \,(\text{CO}_2) + \Delta_f H^\circ \,(\text{C(s, graphite)})\right)$$

$$\Delta_r H^\circ = 2 \cdot \left(-110.62 \, \frac{\text{kJ}}{\text{mol}}\right) - \left(\left(-393.77 \, \frac{\text{kJ}}{\text{mol}}\right) + \left(0 \, \frac{\text{kJ}}{\text{mol}}\right)\right) = 172.53 \, \frac{\text{kJ}}{\text{mol}}$$

$$\Delta_r S^\circ = S^\circ \,(\text{Products}) - S^\circ \,(\text{Reactants})$$

$$\Delta_r S^\circ = 2 \cdot S^\circ \,(\text{CO}) - (S^\circ \,(\text{CO}_2) + S^\circ \,(\text{C(s, graphite)}))$$

$$\Delta_r S^\circ = 2 \cdot \left(198.12 \, \frac{\text{J}}{\text{mol K}}\right) - \left(\left(213.86 \, \frac{\text{J}}{\text{mol K}}\right) + \left(5.74 \, \frac{\text{J}}{\text{mol K}}\right)\right) =$$

$$+ 176.64 \, \frac{\text{J}}{\text{mol K}}$$

$$\Delta_r G^\circ = \Delta_r H^\circ - T\Delta_r S^\circ$$

$$\Delta_r G^\circ = \left(-172.53\,\frac{kJ}{mol}\right) - 798.15\ K\ \left(+176.64\,\frac{J}{mol\ K}\right) = +35.96\,\frac{kJ}{mol}$$

$$\ln\{K_{eq}\} = -\frac{\Delta_r G^\circ}{RT} = -\frac{35960\,\frac{J}{mol\ k}}{8.314\,\frac{J}{mol\ K}\ 773.15\ K} = -5.59$$

$$\{K_{eq}\} = e^{-5.59} = 0.0037$$

$$K_{eq} = \frac{[CO]^2_{eq}}{[CO_2]_{eq}[C]_{eq}}$$

$$[K_{eq}] = \frac{bar^2}{bar \cdot \frac{mol}{mol}} = bar$$

$$K_{eq} = 0.0037\ bar$$

Chapter 6 "Vapor Pressure"

Solutions to the Test Questions

1. a: Vapor pressure increases
   b: Vapor pressure remains constant
      c: Vapor pressure increases very slightly
      d: Vapor pressure decreases
2. d
3. a and b

Solutions to the Exercises

1.

$$\frac{p_{O_2}}{x_{O_2}} = K_{\text{Henry}}$$

$$x_{O_2} = \frac{p_{O_2}}{K_{\text{Henry}}} = \frac{21 \text{ kPa}}{4.6 \text{ GPa}} = 4.6 \cdot 10^{-6} \frac{\text{mol}}{\text{mol}} \, (= 4.6 \text{ mol} - \text{ppm})$$

$$x_{O_2} = \frac{n_{O_2}}{n_{O_2} + n_{H_2O}} \approx \frac{n_{O_2}}{n_{H_2O}}$$

$$n_{H_2O} = \frac{m}{M} = \frac{1000 \text{g}}{18 \frac{\text{g}}{\text{mol}}} = 56 \text{ mol}$$

$$n_{O_2} = n_{H_2O} \cdot x_{O_2} = 4.6 \cdot 10^{-6} \frac{\text{mol}}{\text{mol}} \cdot 56 \text{ mol} = 0.25 \text{ mmol (approx.5.7 mL@STP)}$$

2.

$$\log_{10}(p \text{ in kPa}) = A - \frac{B}{C + (T \text{ in } °C)}$$

	A	B	C
CH_3COCH_3	6.24204	1210.595	229.664

$$\log_{10}(72) = 6.24204 - \frac{1210.595}{229.664\,°C + T}$$

$$T = 46.4\,°C$$

3.

$$\log_{10}(p \text{ in kPa}) = A - \frac{B}{C + (T \text{ in } °C)}$$

	A	B	C
C_2H_5OH	7.2371	1592.864	226.184

$$\log_{10}(50) = 7.2371 - \frac{1592.864}{226.184\,°C + T}$$

$$T = 61.4\,°C$$

4.

$$\ln\left(\frac{p_2^*}{p_1^*}\right) = -\frac{\Delta_{vap}H}{R}\left(\frac{1}{T_2} - \frac{1}{T_1}\right)$$

a

$$\ln\left(\frac{0.743 \text{ kPa}}{1.30 \text{ kPa}}\right) = -\frac{\Delta_{vap}H}{8.314\,\frac{J}{K\,mol}}\left(\frac{1}{276.88 \text{ K}} - \frac{1}{283.74 \text{ K}}\right)$$

$$\Delta_{vap}H = 53.3\,\frac{kJ}{mol}$$

b

$$\ln\left(\frac{p_2^*}{p_1^*}\right) = -\frac{\Delta_{vap}H}{R}\left(\frac{1}{T_2} - \frac{1}{T_1}\right)$$

$$\ln\left(\frac{100 \text{ kPa}}{1.30 \text{ kPa}}\right) = -\frac{53265\,\frac{J}{mol}}{8.314\,\frac{J}{K\,mol}}\left(\frac{1}{T} - \frac{1}{283.74 \text{ K}}\right)$$

$$T = 351 \text{ K } (78\,°\text{C})$$

5.

$$\log_{10}(p \text{ in kPa}) = A - \frac{B}{C + (T \text{ in } °\text{C})}$$

	A	B	C
H_2O	7.19621	1730.63	233.426

$$\log_{10}(p^*) = 7.19621 - \frac{1730.63}{233.426\,°\text{C} + 100.0\,°\text{C}}$$

$$p^*(100\,°\text{C}) = 101.3 \text{ kPa}$$

$$p_{H_2O} = \varphi \cdot p^*_{H_2O} = 0.200 \cdot 101.3 \text{ kPa} = 20.3 \text{ kPa}$$

$$\log_{10}(20.3) = 7.19621 - \frac{1730.63}{233.426\,°\text{C} + T}$$

$$T_{\text{dew}} = 60.4\,°\text{C}$$

Chapter 7 "Solutions"

Solutions to the Test Questions

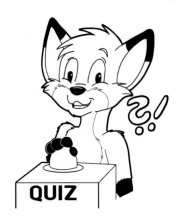

1. d: lowest boiling point
   b: lowest freezing point

2. b
3. b

Solutions to the Exercises

1.

$$\gamma = \frac{m}{V} = \frac{3.50 \text{ mg}}{5.00 \text{ mL}} = 0.700 \frac{\text{g}}{\text{L}} \left(700 \frac{\text{mg}}{\text{L}} \right)$$

$$\Pi = c_B \cdot R \cdot T \cdot i$$

$$0.205 \text{ kPa} = c_B \cdot 8.314 \frac{\text{kPa L}}{\text{mol K}} \cdot 298 \text{ K} \cdot 1$$

$$c_B = 8.27 \cdot 10^{-5} \frac{\text{mol}}{\text{L}} \left(82.7 \frac{\mu \text{mol}}{\text{L}} \right)$$

$$M = \frac{m}{n} = \frac{\gamma}{c} = \frac{0.700 \frac{\text{g}}{\text{L}}}{8.27 \cdot 10^{-5} \frac{\text{mol}}{\text{L}}} = 8460 \frac{\text{g}}{\text{mol}}$$

2.

$$\Pi = c_B \cdot R \cdot T \cdot i$$

$$780 \text{ kPa} = c_B \cdot 8.314 \frac{\text{kPa L}}{\text{mol K}} \cdot 310 \text{ K} \cdot 1$$

$$c_B = 0.303 \frac{\text{mol}}{\text{L}} \left(303 \frac{\text{mmol}}{\text{L}} \text{ oder } 303 \frac{\text{mosmol}}{\text{L}} \right)$$

3.

$$n_{\text{Urea}} = \frac{m}{M} = \frac{60.9 \text{ g}}{60.06 \frac{\text{g}}{\text{mol}}} = 1.014 \text{ mol}$$

$$b = \frac{n_{\text{Urea}}}{m_{\text{Water}}} = \frac{1.014 \text{ mol}}{0.500 \text{ kg}} = 2.028 \frac{\text{mol}}{\text{kg}}$$

$$m_{\text{Solution}} = m_{\text{Water}} + m_{\text{Urea}} = 0.5609 \text{ kg}$$

$$V_{\text{Solution}} = \frac{m_{\text{Solution}}}{\rho_{\text{Solution}}} = \frac{0.5609 \text{ kg}}{1.000 \frac{\text{kg}}{\text{L}}} = 0.5609 \text{ L}$$

$$c = \frac{n_{\text{Harnstoff}}}{V_{\text{Solution}}} = \frac{1.014 \text{ mol}}{0.5609 \text{ L}} = 1.808 \frac{\text{mol}}{\text{L}}$$

$$x_{\text{Urea}} = \frac{n_{\text{Urea}}}{n_{\text{Urea}} + n_{\text{Water}}} = \frac{1.014 \text{ mol}}{1.014 \text{ mol} + 27.778 \text{ mol}} = 0.03522 \ (3.5 \text{ mol} - \%)$$

$$a : \Delta_{\text{fus}} T = - k_{\text{kr}} \cdot b_B \cdot i$$

$$\Delta_{\text{fus}} T = - 1.86 \frac{\text{K kg}}{\text{mol}} \cdot 2.028 \frac{\text{mol}}{\text{kg}} \cdot 1 = - 3.77 \text{ K}$$

$$T_{\text{fus}} = - 3.77 \,^{\circ}\text{C}$$

$$b : \Pi = c_B \cdot R \cdot T \cdot i$$

$$\Pi = 1.808 \frac{\text{mol}}{\text{L}} \cdot 8.314 \frac{\text{kPa L}}{\text{mol K}} \cdot 284.35 \text{ K} \cdot 1 = 4.27 \text{ MPa}$$

$$c : \Delta p = - x_{\text{urea}} \cdot p^* \cdot i$$

$$\Delta p = - 0.03522 \frac{\text{mol}}{\text{mol}} \cdot 101.325 \text{kPa} \cdot 1 = - 3.57 \text{ kPa}$$

$$p = 97.8 \text{ kPa}$$

4.

$$n_{NaCl} = \frac{m}{M} = \frac{11.23 \text{ g}}{58.44 \frac{\text{g}}{\text{mol}}} = 0.1922 \text{ mol}$$

$$a : c = \frac{n_{NaCl}}{V_{L\ddot{o}sung}} = \frac{0.1922 \text{ mol}}{1.00 \text{ L}} = 0.192 \frac{\text{mol}}{\text{L}} \left(= 192 \frac{\text{mmol}}{\text{L}} \right)$$

$$b : i \cdot c = 2 \cdot 0.192 \frac{\text{mol}}{\text{L}} = 0.384 \frac{\text{mol}}{\text{L}} \left(= 384 \frac{\text{mmol}}{\text{L}} = 384 \frac{\text{mosmol}}{\text{L}} \right)$$

$$c : \Pi = c_B \cdot R \cdot T \cdot i$$

$$\Pi = 0.192 \frac{\text{mol}}{\text{L}} \cdot 8.314 \frac{\text{kPa L}}{\text{mol K}} \cdot 307.45 \text{ K} \cdot 1 = 982 \text{ kPa}$$

Chapter 8 "Phase Diagrams"

Solutions to the Test Questions

1. c
2. a, b, c, and d

3. Eutectic at approx. $-20\ °C$ and 80 mol% water
   Peritectic at approx. $0\ °C$ and 66 mol% water (stoichiometric compound halite)

Solutions to the Exercises

1.
(a) At 632 K, the mixture starts to melt; the eutectic melt is formed (approx. 65 % LiCl)
(b) At approx. 750 K, the melt starts to solidify; the resulting crystals are (almost) pure KCl
2.
   The three-component system is heterogeneous and consists of 11 kg organic phase (toluene) and 9 kg aqueous phase (mixture of 7 kg water and 2 kg acetic acid)

Chapter 9 "Reaction Kinetics"

Solutions to the Test Questions

(Table 13.1)

1. a and e
2. a: $r°$ doubles; $t°_{1/2}$ remains constant
   b: $r°$ quadruples; $t°_{1/2}$ halves
3. b and e

Table 13.1 Kinetic data for simple reactions

Reaction	Order	Rate law	$[k]$	Integrated rate law	Half-life
$A \rightarrow P$	0	$r = k$	$\frac{mol}{L\ s}$	$[A] = [A]_0 - kt$	$t_{1/2} = \frac{[A]_0}{2\,k}$
$A \rightarrow P$	1	$r = k[A]$	$\frac{1}{s}$	$[A] = [A]_0 \cdot e^{-kt}$	$t_{1/2} = \frac{ln(2)}{k}$
$A \rightarrow P$	2	$r = k$ $[A]^2$	$\frac{L}{mol\ s}$	$[A] = \frac{[A]_0}{1+k[A]_0 t}$	$t_{1/2} = \frac{1}{k\,[A]_0}$
$A + B \rightarrow P$	2 (1 + 1)	$r = k[A]$ $[B]$	$\frac{L}{mol\ s}$	$kt = \frac{1}{[B]_0 - [A]_0}\ ln\left(\frac{[B][A]_0}{[A][B]_0}\right)$	Depending on stoichiometry

Solutions to the Exercises

1.

$$[A] = \frac{[A]_0}{1 + k[A]_0 t} = \frac{0.0500\,\frac{mol}{L}}{1 + 0.00985\,\frac{L}{mol\ s}\ 0.0500\,\frac{mol}{L}\ 1800\ s} = 0.0265\,\frac{mol}{L}$$

$$\text{Umsatz} = \frac{[A]_0 - [A]}{[A]_0} \cdot 100\% = \frac{0.0500\,\frac{mol}{L} - 0.0265\,\frac{mol}{L}}{0.0500\,\frac{mol}{L}} = 45.0\%$$

$$t_{1/2} = \frac{1}{k\,[A]_0} = \frac{1}{0.00985\,\frac{L}{mol\ s}0.0500\,\frac{mol}{L}} = 2030\ s = 33.8\ min$$

$$r = k[A]^2 = r = 0.00985\,\frac{L}{mol\ s}\left[0.0500\,\frac{mol}{L}\right]^2 = 2.46 \cdot 10^{-5}\,\frac{mol}{L\ s} = 88.7\,\frac{mmol}{L\ h}$$

2.

$$\ln\left(\frac{k'(T_1)}{k(T_2)}\right) = -\frac{E_A}{R}\left(\frac{1}{T_1} - \frac{1}{T_2}\right)$$

$$E_A = -R\frac{\ln\left(\frac{k'(T_1)}{k(T_2)}\right)}{\left(\frac{1}{T_1} - \frac{1}{T_2}\right)} = -R\frac{\ln\left(\frac{t_{1/2}(T_2)}{t'_{1/2}(T_1)}\right)}{\left(\frac{1}{T_1} - \frac{1}{T_2}\right)}$$

$$E_A = -8.314 \frac{J}{mol\ K} \cdot \frac{\ln\left(\frac{2.90\ min}{10.0\ min}\right)}{\left(\frac{1}{303.15\ K} - \frac{1}{323.15\ K}\right)} = 50.4 \frac{kJ}{mol}$$

3.

$$a = 1; b = 0; k = 0.40 \frac{1}{s}$$

Chapter 10 "Reaction Mechanism"

Solutions to the Test Questions

1. a and d
2. Consecutive reaction: slowest step
   Parallel reaction: fastest step
3. a, c, and d
4. b and c

Solutions to the Exercises

1.

$$\Delta_R H = \overrightarrow{E_A} - \overleftarrow{E_A} = 11.9\,\frac{kJ}{mol} - 19.4\,\frac{kJ}{mol} = -7.5\,\frac{kJ}{mol}$$

$$\ln\left(\frac{k'(T_1)}{k(T_2)}\right) = -\frac{E_A}{R}\left(\frac{1}{T_1} - \frac{1}{T_2}\right)$$

$$\ln\left(\frac{k'(313.7\ \mathrm{K})}{k(297.2\ \mathrm{K})}\right) = -\frac{11900\,\frac{J}{mol}}{8.314\,\frac{J}{mol\ K}}\left(\frac{1}{313.7\ \mathrm{K}} - \frac{1}{297.2\ \mathrm{K}}\right)$$

$$\ln\left(\frac{k'(313.7\ \mathrm{K})}{11.9\frac{1}{h}}\right) = -\frac{11900\,\frac{J}{mol}}{8.314\,\frac{J}{mol\ K}}\left(\frac{1}{313.7\ \mathrm{K}} - \frac{1}{297.2\ \mathrm{K}}\right) = 0.253$$

$$k'(313.7\ \mathrm{K}) = 11.9\frac{1}{h}\cdot e^{0.253} = 15.3\frac{1}{h}$$

2. a: kinetic product B: low temperature, short reaction time, catalyst
   b: thermodynamic product C: high temperature, long reaction time

Chapter 11 "Electrolytes"

Solutions to the Test Questions

1. a > b > c > d > e > f
2. c
3. b, c, and d

Solutions to the Exercises

1.

$$E = \frac{U}{d} = \frac{60.8 \text{ V}}{0.248 \text{ m}} = 245 \frac{\text{V}}{\text{m}}$$

$$LiCl \rightarrow Li^+ + Cl^-$$

$$u_+ = \frac{\lambda_+}{F} = \frac{3.87 \frac{mS\ m^2}{mol}}{96485 \frac{A\ s}{mol}} = 4.01 \cdot 10^{-8} \frac{m^2}{V\ s}$$

$$v_+ = u_+ \cdot E = 4.01 \cdot 10^{-8} \frac{m^2}{V\ s} \cdot 245 \frac{V}{m} = 9.83 \cdot 10^{-6} \frac{m}{s} \left(= 3.5 \frac{cm}{h}\right)$$

$$t_+ = \frac{v_+ \lambda_+}{v_+ \lambda_+ + v_- \lambda_-} = \frac{1 \cdot 3.87 \frac{mS\ m^2}{mol}}{1 \cdot 3.87 \frac{mS\ m^2}{mol} + 1 \cdot 7.63 \frac{mS\ m^2}{mol}} = 0.337\ (34\%)$$

2.

a

$$H_2SO_4 \rightarrow 2\ H^+ + SO_4^{2-}$$

$$c = \frac{n}{V} = \frac{1.00\ mmol}{1.00\ L} = 1.00 \cdot 10^{-3} \frac{mol}{L} = 1.00 \frac{mol}{m^3}$$

$$\Lambda_\infty = v^+ \lambda_\infty^+ + v^- \lambda_\infty^-$$

$$\Lambda_\infty = 2 \cdot 34.96 \frac{mS\ m^2}{mol} + 1 \cdot 16.0 \frac{mS\ m^2}{mol} = 85.92 \frac{mS\ m^2}{mol}$$

$$\Lambda = \Lambda_\infty - K\sqrt{c} = 85.92 \frac{mS\ m^2}{mol} - 364 \frac{mS\ m^2}{mol\sqrt{\frac{mol}{L}}} \sqrt{0.001 \frac{mol}{L}} = 74.42 \frac{mS\ m^2}{mol}$$

$$\kappa = \Lambda \cdot c$$

$$\kappa = 74.42 \frac{mS\ m2}{mol} \cdot 1.00 \frac{mol}{m^3} = 74.4 \frac{mS}{m} \left(= 744 \frac{\mu S}{cm}\right)$$

$$[H^+] = \alpha \cdot v_{H^+} \cdot c = 1 \cdot 2 \cdot 1.00 \cdot 10^{-3} \frac{mol}{L} = 2.00 \cdot 10^{-3} \frac{mol}{L}$$

$$pH = - \log\left(\frac{[H^+]}{mol/L}\right) = - \log(0.00200) = 2.7$$

$$HOAc \rightarrow H^+ + OAc^-$$

$$c = \frac{n}{V} = \frac{1.00 \text{ mmol}}{1.00 \text{ L}} = 1.00 \cdot 10^{-3} \frac{\text{mol}}{\text{L}} = 1.00 \frac{\text{mol}}{\text{m}^3}$$

$$\Lambda_\infty = \nu^+ \lambda_\infty^+ + \nu^- \lambda_\infty^-$$

$$\Lambda_\infty = 1 \cdot 34.96 \frac{\text{mS m}^2}{\text{mol}} + 1 \cdot 4.09 \frac{\text{mS m}^2}{\text{mol}} = 39.05 \frac{\text{mS m}^2}{\text{mol}}$$

$$\Lambda = \alpha \Lambda_\infty$$

$$\frac{\alpha^2}{1 - \alpha} c = K_a$$

$$\frac{\alpha^2}{1 - \alpha} 1.00 \cdot 10^{-3} \frac{\text{mol}}{\text{L}} = 1.3 \cdot 10^{-5} \frac{\text{mol}}{\text{L}}$$

$$\alpha = 0.108 \ (10.8\%)$$

$$\Lambda = \alpha \Lambda_\infty = 0.108 \cdot 39.05 \frac{\text{mS m}^2}{\text{mol}} = 4.22 \frac{\text{mS m}^2}{\text{mol}}$$

$$\kappa = \Lambda \cdot c$$

$$\kappa = 4.22 \frac{\text{mS m}^2}{\text{mol}} \cdot 1.00 \frac{\text{mol}}{\text{m}^3} = 4.22 \frac{\text{mS}}{\text{m}} \left(= 42.2 \frac{\mu\text{S}}{\text{cm}} \right)$$

$$[\text{H}^+] = \alpha \cdot \nu_{\text{H}^+} \cdot c = 0.108 \cdot 1 \cdot 1.00 \cdot 10^{-3} \frac{\text{mol}}{\text{L}} = 1.08 \cdot 10^{-4} \frac{\text{mol}}{\text{L}}$$

$$\text{pH} = - \log \left(\frac{[\text{H}^+]}{\frac{\text{mol}}{\text{L}}} \right) = - \log(0.000108) = 3.97$$

Chapter 12 "Electrodes"

Solutions to the Test Questions

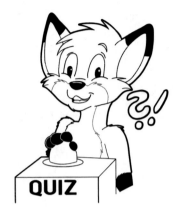

1. b, d, and f
2. a
3. a and c

Solutions to the Exercises

1.

$$Zn^{2+}(aq) + 2\,e^- \rightleftharpoons Zn\,(s)$$

$$n = \frac{2.81 \text{ g}}{65.38 \frac{\text{g}}{\text{mol}}} = 0.0430 \text{ mol}$$

$$I \cdot t = n \cdot \nu_e \cdot F = 0.0430 \text{ mol} \cdot 2 \cdot 96485 \frac{\text{As}}{\text{mol}} = 8.29 \text{ kAs} \ (2300 \text{ mAh})$$

$$w_{el} = U \cdot I \cdot t = 0.99 \text{ V} \cdot 8294 \text{ A s} = 8.21 \text{ kJ} \ (8211 \text{ Ws} = 0.0022 \text{ kWh})$$

2.

$$H_2(g) + \frac{1}{2}O_2(g) \rightarrow H_2O \ (l)$$

$$\Delta_r H^\circ = \Delta_f H^\circ \left(H_2O \ (l)\right) - \left(\Delta_f H^\circ \left(H_2 \ (g)\right) + \frac{1}{2}\Delta_f H^\circ \left(O_2 \ (g)\right)\right) = -285.8 \frac{\text{kJ}}{\text{mol}}$$

$$\Delta_r G^\circ = \Delta_f G^\circ \left(H_2O \ (l)\right) - \left(\Delta_f G^\circ \left(H_2 \ (g)\right) + \frac{1}{2}\Delta_f G^\circ \left(O_2 \ (g)\right)\right) = -237.1 \frac{\text{kJ}}{\text{mol}}$$

$$\Delta_r S^\circ = S^\circ \left(H_2O \ (l)\right) - \left(S^\circ \left(H_2 \ (g)\right) + \frac{1}{2}S^\circ \left(O_2 \ (g)\right)\right) = -163.3 \frac{\text{J}}{\text{K mol}}$$

$$w_{el.rev} = n \cdot \Delta_r G^\circ = 1 \text{ mol} \cdot \left(-237.1 \frac{\text{kJ}}{\text{mol}}\right) = -237.1 \text{ kJ} \ (= -0.066 \text{ kWh})$$

$$q_{rev} = n \cdot T \cdot \Delta_r S^\circ = 1 \text{ mol} \cdot 298 \text{ K} \cdot \left(-163.3 \frac{\text{kJ}}{\text{mol}}\right) =$$
$$-48.7 \text{ kJ} \ (= -0.014 \text{ kWh})$$

$$\eta = \frac{\Delta_r G^\circ}{\Delta_r H^\circ} = \frac{-237.1 \frac{\text{kJ}}{\text{mol}}}{-285.8 \frac{\text{kJ}}{\text{mol}}} = 0.829 \ (83\%)$$

3.

$$\Delta_{\text{Mem}}\varphi = \varphi(\text{II}) - \varphi(\text{I}) = -\frac{RT}{z_i F} \ln \frac{[i]^{\text{II}}}{[i]^{\text{I}}}$$

$$\Delta_{\text{Mem}}\varphi = \frac{8.314 \frac{\text{VAs}}{\text{mol K}} \ 310 \text{ K}}{1 \ 96485 \frac{\text{As}}{\text{mol}}} \ln \left(\frac{155 \frac{\text{mmol}}{\text{L}}}{4.00 \frac{\text{mmol}}{\text{L}}}\right) = -0.098 \text{ V} \ (-98 \text{ mV})$$

Intracellular fluid is negatively charged

4.

$$\text{Red}$$
$$4\,\text{H}^+\,(\text{aq}) + 4\,e^- + O_2(g) \rightleftarrows 2\,H_2O(l)$$
$$\text{Ox}$$

$$E_{H_2O/O_2} = E^\circ{}_{H_2O/O_2} + \frac{RT}{4F}\ln\frac{[H^+]^4 \cdot [O_2]}{[H_2O]^2}$$

$$pH = 7 \Rightarrow [H^+] = 10^{-7}\left(\frac{mol}{L}\right)$$

$$p_{O_2} = 100\ kPa \Rightarrow [O_2] = 1.00\ (bar)$$

$$E_{H_2O/O_2} = 1.23\ V + \frac{8.314\,\frac{J}{mol\ K} \cdot 298.15\ K}{4 \cdot 96485\,\frac{As}{mol}}\ln\frac{\left(10^{-7}\right)^4 \cdot [1]}{[1]^2} = 0.82\ V$$

$$\text{Red}$$
$$\text{Ag}^+(\text{aq}) + e^- \rightleftarrows \text{Ag\,(s)}$$
$$\text{Ox}$$

$$E_{Ag/Ag^+} = E^\circ{}_{Ag/Ag^+} + \frac{RT}{1\,F}\ln\frac{[Ag^+]}{[Ag]}$$

$$c_{Ag^+} = 10^{-4}\frac{mol}{L} \Rightarrow [Ag^+] = 10^{-4}\left(\frac{mol}{L}\right)$$

$$E_{Ag/Ag^+} = 0.80\ V + \frac{8.314\,\frac{J}{mol\ K} \cdot 298\ K}{96485\,\frac{As}{mol}}\ln\frac{0.00100}{1} = 0.56\ V$$

$$E_{H_2O/O_2} > E_{Ag/Ag^+}$$

$$E_{H_2O/O_2} : \text{Cathode; negative terminal}$$

$$E_{Ag/Ag^+} : \text{Anode; positive terminal}$$

$$\Delta E = E_{cathode} - E_{anode} = E_{H_2O/O_2} - E_{Ag/Ag^+} = 0.82\ V - 0.56\ V = 0.26\ V$$

13.2   Classical Lab Experiments in Physical Chemistry (Fig. 13.2)

The following experiments have been standard in all basic physical chemistry courses for many decades and are well suited to supplement the workshops.

Detailed experiment instructions can be found in the multimedia lab.

The Conductivity of Strong and Weak Electrolytes (Fig. 13.3)

Theory
Learn (e.g., in Chap. 11) about the electrical conductivity of electrolytes.

Task
- Measure the conductivity of a weak electrolyte (acetic acid) and a strong one (NaCl) at six different concentrations each.
- Plot the molar conductivities of both electrolyte solutions against $\sqrt{c}$ and determine Λ_∞ (NaCl) graphically.
- Plot $\frac{1}{\Lambda}$ against $(\Lambda \cdot c)$ and determine Λ_∞ and K_a for acetic acid.

Fig. 13.2 Multimedia practical PhysChemBasicslight (https://www.ili.fh-aachen.de/goto_elearning_grp_584340.html)

Fig. 13.3 Apparatus for conductometric measurements on electrolytes (https://doi.org/10.5446/59216)

Preparation Questions
- What are the units of specific conductivity κ and molar conductivity Λ?
- The limiting molar conductivity of sodium acetate Λ_∞ (NaOAc) at 25 °C is 9.09 mS m^2/mol. From this value and the table values of acetic acid and saline solution, determine the limiting conductivity of hydrochloric acid Λ_∞ (HCl).
- Calculate the specific conductivity κ (H$_2$O) of pure water at pH 7.

Evaluation
Strong Electrolytes

In contrast to weak electrolytes, strong electrolytes are always completely dissociated in aqueous solution. The concentration dependence of the molar conductivity is given by KOHLRAUSCH's square root law at a given temperature

$$\Lambda = \Lambda_\infty - K_K \sqrt{c}$$

Λ_∞ is called the limiting molar conductivity, it corresponds to the molar conductivity at infinite dilution ($c \to 0$). In the state of infinite dilution, interactions between ions no longer exist. If we plot Λ against $\sqrt{c}$, we obtain a straight line whose ordinate intercept corresponds to the limiting molar conductivity Λ_∞ (Fig. 13.4).

Weak Electrolytes

According to the law of mass action, the dissociation constant K of acetic acid (HOAc), a typical weak electrolyte:

$$K_a = \frac{[\text{H}^+]_{\text{eq}} \cdot [\text{CH}_3\text{COO}^-]_{\text{eq}}}{[\text{CH}_3\text{COOH}]_{\text{eq}}}$$

The degree of dissociation α OSTWALD's dilution law is then obtained:

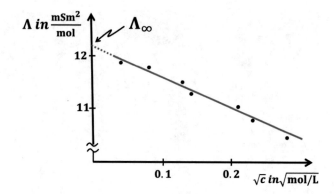

Fig. 13.4 Determination of the limiting molar conductivity of saline solution according to KOHLRAUSCH

$$K_a = \frac{\alpha^2}{1 - \alpha} \, c$$

At infinite dilution, even a weak electrolyte is completely dissociated. Thus, the limiting molar conductivity Λ_∞ of a weak electrolyte corresponds to complete dissociation ($\alpha = 1$). Assuming that the sharp decrease in the molar conductivity of weak electrolytes with increasing concentration is due solely to the decrease in the degree of dissociation, we can formulate:

$$\Lambda = \alpha \cdot \Lambda_\infty$$

Transformation results in:

$$\frac{1}{\Lambda} = \frac{1}{\Lambda_\infty} + \frac{1}{K_a \left(\Lambda_\infty\right)^2} \cdot \Lambda \, c$$

This equation has the form of a straight line equation. Plotting $\frac{1}{\Lambda}$ against $\Lambda \, c$ leads to a straight line, which can be evaluated for Λ_∞ and K_a (Fig. 13.5).

The Kinetics of Cane Sugar Inversion (Fig. 13.6)

Theory
Review the basics of the kinetics of simple reactions (Chap. 9).

Task
- Determine the rate constant for the acid-catalyzed cleavage of cane sugar into glucose and fructose at approx. 30 °C. To determine the rate constant k, we plot ln

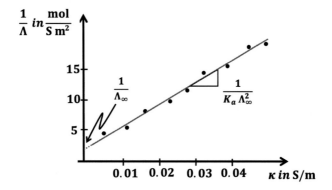

Fig. 13.5 Determination of the limiting molar conductivity of acetic acid according to OSTWALD

($\alpha - \alpha\infty$) on the ordinate and the corresponding times t on the abscissa and determine the rate constant from the slope of this straight line.
- Plot your pair of measurements together with literature data in an ARRHENIUS plot and determine the activation energy graphically.

Preparation Questions
- What are the units of the reaction rate r and the rate constant k?
- Draw the reaction profile of an exothermic reaction. Label the activation energy and the enthalpy of reaction in the diagram.
- Calculate the half-life of the cleavage reaction at 70 °C.
- In a first-order reaction, the initial concentrations of all reactants involved are doubled. How does the initial reaction rate and the half-life of the reaction change?

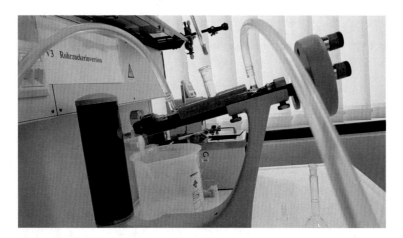

Fig.13.6 Apparatus for polarimetric measurement of the cleavage of sucrose into glucose and fructose ("Sucrose Inversion") (https://doi.org/10.5446/59218)

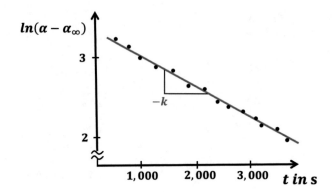

Fig. 13.7 Determination of the rate constant of cane sugar inversion

Evaluation
The course of the sugar inversion can be followed well, since both the cane sugar and the invert sugar solution rotate the plane of oscillation of polarized light. Accordingly, the integrated rate law can also be formulated with the rotation angles (Fig. 13.7)

$$\ln\left(\frac{\alpha_0 - \alpha_\infty}{\alpha - \alpha_\infty}\right) = k \cdot t$$

$$\ln(\alpha - \alpha_\infty) = \ln(\alpha_0 - \alpha_\infty) - k \cdot t$$

To determine the activation energy, we need to measure the rate constant at several temperatures. Then we can evaluate the measurements graphically (ARRHENIUS plot) (Fig. 13.8).

Vapor Pressure Curve and Enthalpy of Evaporation (Fig. 13.9)

Theory
Learn about the basics of phase equilibria (Chap. 6).

Task
- Measure the vapor pressure of an organic liquid as a function of temperature and graphically determine the enthalpy of vaporization $\Delta_{vap}H°$.
- Estimate the boiling point of the liquid at 100 kPa (14.5 psi).

Preparation Questions
- Explain the phase diagram of a pure substance (pT diagram). Name the single-phase realms in the diagram (Fig. 13.10).

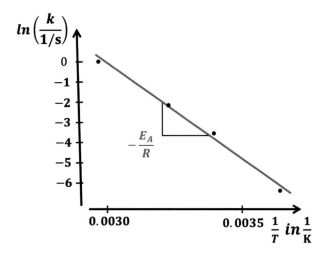

Fig. 13.8 Determination of the activation energy of cane sugar inversion according to Arrhenius

- A liquid with the (constant) enthalpy of vaporization 40.0 kJ/mol boils at 100 °C at a pressure of 100 kPa. Determine the boiling point of the liquid at a pressure of 1.00 MPa.

Evaluation

In this experiment, we are interested in the liquid/gas phase boundary, the vapor pressure curve (Fig. 13.11).

For the liquid the CLAUSIUS–CLAPEYRON equation is valid

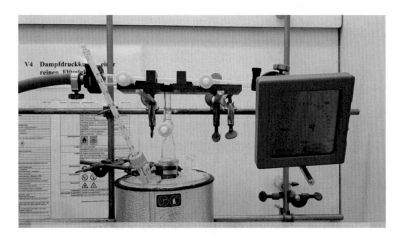

Fig. 13.9 Apparatus for measuring the vapor pressure of methanol at various temperatures (https://doi.org/10.5446/59219)

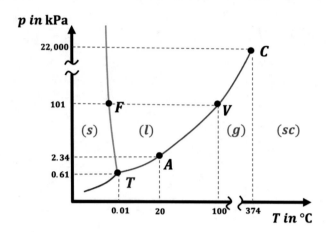

Fig. 13.10 Phase diagram of water

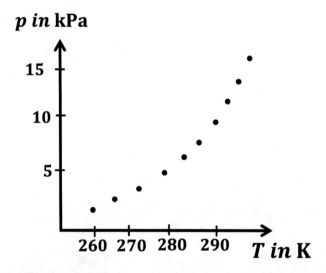

Fig. 13.11 Experimentally determined vapor pressure curve of methanol

$$\ln\left(\frac{p^{*\prime}}{p^{*}}\right) = -\frac{\Delta_{vap}H^{\circ}}{R}\left(\frac{1}{T'} - \frac{1}{T}\right)$$

The average molar enthalpy of vaporization can be calculated using the slope of the plot $\ln\left(\frac{p^{*}}{kPa}\right)$ vs. $\left(\frac{1}{T}\right)$ (Fig. 13.12).

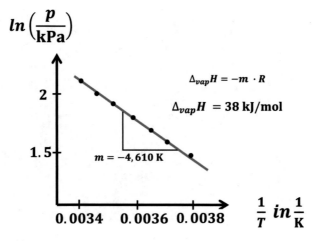

Fig. 13.12 Determination of the enthalpy of vaporization of methanol according to CLAUSIUS– CLAPEYRON

Determination of the Neutralization Enthalpy (Fig. 13.13)

Theory
Learn about the basics of thermochemistry (Chap. 4).

Task
- Determine the calorimeter constant of the measurement setup.
- Determine the molar enthalpy of dilution of hydrochloric acid
- Determine the molar enthalpy of neutralization of sodium hydroxide solution with hydrochloric acid.
- Determine the molar enthalpy of neutralization of ammonia solution with hydrochloric acid.

Preparation Questions
- What is the unit of the calorimeter constant?
- Calculate a value for the enthalpy of neutralization from the tabulated standard enthalpies of formation of the ions H^+ and OH^-.
- A heating filament (power: 100 watts) is immersed in 1 liter of water of initial temperature 20 °C. Current flows through the filament for 10 minutes. Calculate the heat that the heating filament gives off to the water and the final temperature of the water.

Evaluation
The heat capacity of the entire experimental setup C (calorimeter constant) is determined experimentally by filling water into the calorimeter instead of the

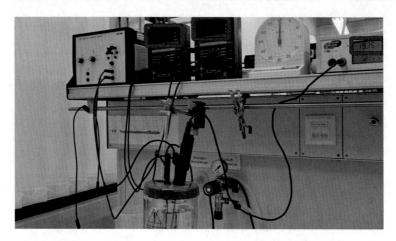

Fig. 13.13 Apparatus for calorimetric determination of neutralization enthalpy (https://doi.org/10.5446/59220)

reaction mixture, and adding a known amount of heat to the system (electrical heating) and determining ΔT. It is assumed that there is no difference in heat capacity between pure water and dilute aqueous solutions (Fig. 13.14).

We have to subtract the enthalpy of dilution from the measured enthalpy change of the reactions between alkali and acid to obtain the actual enthalpy of neutralization. The enthalpy of dilution is determined in a separate experiment, where concentrated hydrochloric acid is mixed with pure water (Figs. 13.15 and 13.16).

Migration of Ions in an Electric Field (Fig. 13.17)

Theory
Learn about the behavior of electrolytes (Chap. 11).

Task
- Determine the drift velocity of the MnO_4^- ion at two different electric fields. Plot the migration distance against time and determine the slope of the straight line. MnO_4^-.
- From the drift velocities, calculate the ionic mobility u_-, the ionic conductivity λ_- and the radius of the hydrated MnO_4^- ion r_-.

Preparation Questions
- What are the units of ionic mobility and ionic molar conductivity?
- A potassium permanganate solution is electrolyzed. The voltage is 2 V; the distance between the electrodes is 5 cm. Determine the velocity of the potassium ion and the permanganate ion in the electric field.

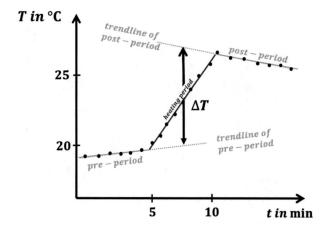

Fig. 13.14 Determining the temperature rise for calculating the calorimeter constant

Fig. 13.15 Determination of
the enthalpy of neutralization
from the enthalpy of dilution
according to Hess' theorem

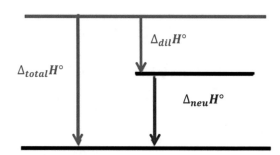

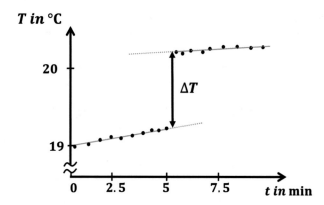

Fig. 13.16 Determining the temperature rise for calculating the enthalpy of dilution

Fig. 13.17 Apparatus for measuring the drift velocity of permanganate ions (https://doi.org/10.5446/59221)

$$\lambda_-^{\infty}\left(MnO_4^-\right) = 6.13 \,\frac{mSm^2}{mol} \quad \lambda_+^{\infty}(K^+) = 7.34 \,\frac{mSm^2}{mol}$$

Evaluation

The drift velocity v of an ion can be measured directly if we carefully superimpose two electrolyte solutions, each containing a common and a differently colored ion species, and follow the migration of the color boundary under the influence of an electric field. In this experiment, the electrolytes KNO_3 and $KMnO_4$ are used. The drift velocity of the violet colored permanganate ion (MnO_4^-) is measured (Fig. 13.18).

The drift velocity v is related to the molar ionic conductivity

Determination of Molar Mass by Cryoscopy (Fig. 13.19)

Theory

Learn about the behavior of solutions (Chap. 7)

Task

- Determine the molar mass of an unknown substance from the decrease in freezing point at various mass concentrations.

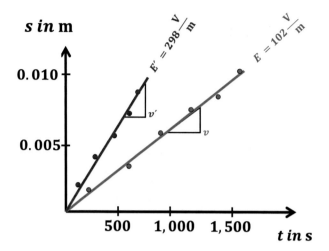

Fig. 13.18 Determination of the drift velocity in two different electric fields

Fig. 13.19 Apparatus for measuring the solidification point of solutions (https://doi.org/10.5446/59217)

Preparation Questions
- What do you mean by colligative properties?
- How many grams of cane sugar ($M = 342.30$ g/mol) or table salt ($M = 58.44$ g/mol) must we add to 1.00 kg of water to lower the freezing point by 1.0 °C?

Evaluation
The freezing point depression T of a solution with respect to the pure solvent is given for ideal solutions by:

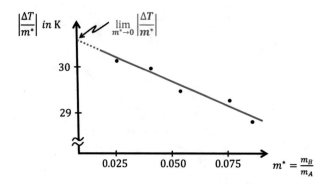

Fig. 13.20 Extrapolation of the measured values to infinite dilution for the determination of the molar mass according to RAOULT's 2nd law

$$\Delta_{\text{fus}}T = -\frac{RT_{\text{fus},A}^2 \, M_A}{\Delta_{\text{fus},A}H} \frac{m_B^*}{M_B} i$$

The substance-specific constants of the solvent and the gas constant can be combined to form the cryoscopic constant k_k:

$$k_k = \frac{RT_{\text{fus},A}^2 \, M_A}{\Delta_{\text{fus},A}H}$$

For water, $k_k = 1.86 \frac{\text{K kg}}{\text{mol}}$

Thus, we obtain for the freezing point depression $\Delta_{\text{fus}}T$:

$$\Delta_{\text{fus}}T = -k_k \frac{m_B^*}{M_B} i$$

This equation is strictly valid only for ideal solutions; for real solutions, we measure the freezing point depression at different mass concentrations of the solute and extrapolate the quotient $\frac{\Delta T}{m_2^*}$ to infinite dilution (Fig. 13.20).

$$\lim_{m_B^* \to 0} \left| \frac{\Delta_{\text{fus}}T}{m_B^*} \right| = \frac{k_k}{M_B} i$$

13.3   Suggestions for Workshop Designs

In the workshop, we deliberately do not use presentation programs (PowerPoint, etc.), but consistently rely on blackboard & chalk (or marker & whiteboard). Important facts are worked out together with the students on the blackboard, visualized, structured with color. Students should be able to trace the key illustrations for each chapter themselves.

The workshops are very experiment-oriented. Experience has shown that the exhibits/experiments, which are often conducted and evaluated by the students themselves, are very well remembered.

A "question of the day" (which is also worked on by the students themselves) helps to provide a common thread through the approx. 90-minute workshop.

Workshop 1: Changes of State

Question of the day
- Does a water bubbler gas cylinder contain liquid carbon dioxide?

Exhibits/Experiments
- pVT phase diagram of a single component system (3D model)
- Kettle (How do we calculate sensible heat?)
- Ammeter and cell phone (How do we calculate electrical work?)
- Fuel cell (How do we perform a chemical reaction reversibly?)
- Water bubbler (phase diagram of CO_2, creating dry ice)
- Gas lighter (How do we describe a two-phase system?)
- Helium balloon (How do we describe a single-phase system?)

Workshop 2: Gases

Question of the day
- How much gaseous water is in the air of the lecture hall?

Exhibits/Experiments
- Hygrometer (What is the relative humidity of the air in the lecture hall?)
- Air pump (What is the compressibility of gases?)
- Radiometer (Why does a light mill turn and why only in one direction?)
- Fire pump (How does a pneumatic lighter work?)
- Methanol gun (How much mL of methanol is needed for stoichiometric combustion in a chip package?)
- Balloon in a vacuum (How does the volume of a balloon change when it rises in the atmosphere to the stratosphere?)
- Table tennis balls (diameter 4 cm) as a model for gas particles; true-to-scale representation of the mean distance (50 cm) and the mean free path (10 m).

Lab Experiments
- Experiment on gas laws (BOYLE–MARIOTTE, GAY-LUSSAC)

Workshop 3: Physical Equilibria

Question of the day
- What is the efficiency of an "ice engine" (STIRLING engine on ice)?

Exhibits/Experiments
- Heatpipe (How fast is the heat transported by the special heatpipe mechanism?)
- STIRLING engine and ice (What happens to the working medium air in a STIRLING engine?)
- Tea in water (How fast does the tea diffuse into the water?)
- Light bulb (How does a thermal conductivity detector (TCD) work?)
- Ammonia and hydrochloric acid (Where do ammonia and hydrochloric acid vapors meet in a glass tube?)
- PELTIER element (Can we generate cold electrically?)

Lab Experiments
- Dissolution rate of gypsum (diffusion-limited)

Workshop 4: Affinity

Question of the day
- Can water burn to hydrogen peroxide?

Exhibits/Experiments
- Slaking lime (how much lime do we need to slake for a hot drink?)
- Dissolving urea (with thermometer; why does an endothermic process occur voluntarily?
- Rubber band (entropy elasticity: how does the temperature change when stretched/relaxed?)
- Bomb calorimeter (How can we practically make an isochoric heat measurement?)
- Thermal power of a tea light oven (Can you heat a room with one of these clay pot heaters?)

Lab Experiments
- Neutralization enthalpy

Workshop 5: Chemical Equilibria

Question of the day
- Does the pH of pure water change with temperature?

Exhibits/Experiments
- Nickel chloride equilibrium (How does the equilibrium change with temperature (color as an indicator of the equilibrium position)?)
- pH meter (How does the pH of pure water change with temperature?)

Workshop 6: Vapor Pressure

Question of the day
- At what temperature does water condense from the air in the lecture hall?

Exhibits/Experiments
- Lemonade bottle with manometer (What is the pressure in a lemonade can?)
- DANIELL dew point hygrometer (How do we determine the dew point by controlled cooling?); hand boiler (How does the vapor pressure change with temperature?); boiling water at negative pressure (syringe)
- EINSTEIN's duck in different modes (How fast does the duck drink when we offer it ethanol instead of water? Why does the duck come to a standstill in a recipient?)
- Swamp cooler (What does "cooling limit temperature" mean?); WILSON fog chamber (How do we generate supersaturated steam?)
- Cavitation: Decrease of pressure in water by impact; cavitation in the animal kingdom and in technology

Lab Experiments
- Vapor pressure curve and enthalpy of vaporization

Workshop 7: Solutions

Question of the day
- At what temperature does seawater freeze and what are the crystals that form made of?

Exhibits/Experiments
- 1M sugar solution and ½ M saline solution (seawater) − (What are isotonic solutions?)
- Chemical garden (silicate solution + metal salt crystals; What is osmosis?)
- Sodium acetate as a latent heat
- Ice fishing (Why does the ice melt when we pour salt on it and then freeze again?)

Lab Experiments
- Molar mass determination by cryoscopy

Workshop 8: Phase Diagrams

Question of the day
- At what temperature does wine boil?

Exhibits/Experiments:
- Kettle and thermometer (We measure the boiling temperature of wine)
- Mechanical lever (What does the lever rule of mechanics say?)
- Adsorption of crystal violet on activated carbon (Is adsorption reversible?)
- Isopropanol/saturated brine − heterogeneous mixture (+ beads for density determination) (The beads are initially right at the phase boundary. Why do some beads float on top after shaking while other beads sink to the bottom)

Lab Experiments
- Boiling point diagram of an ideal two-component system
- Adsorption isotherm activated carbon/acetic acid

Workshop 9: Reaction Kinetics

Question of the day
- What is the thermal power of rusting iron (iron hand warmer)?

Exhibits/Experiments
- Decolorization of crystal violet solution at high pH (How long does it take for complete decolorization?)
- Catalysis of acetone oxidation with copper (Why does the copper "glow by itself?"); spherical path as reaction profile; hydrogen peroxide—decomposition with catalyst
- Iron hand warmer (Fe/air)/Ready-to-eat meal (MRE: Mg/air) (How can an "actually" slow reaction be accelerated?)
- Polarimeter/refractometer/conductometer for concentration measurement (How do we measure the concentration of the reactants without disturbing the reaction?)

Lab Experiments
- Kinetics of cane sugar inversion
- Kinetics of ester cleavage

Workshop 10: Reaction Mechanism

Question of the day
- How much radon is in equilibrium with 1 kg of concrete (30 Bq)?

Exhibits/Experiments
- 3 dropping funnels (analogy of the subsequent reaction—How does the position of the taps influence the level in the middle dropping funnel?)
- Geiger counter (radioactive decay series—What radiation does the Geiger counter indicate? Why is the steel dose higher near green uranium glass?)

Lab Experiments
- Mutarotation of glucose

Workshop 11: Conductivity

Question of the day
- How does dissolving a salt crystal (<1 mg) change the conductivity of water?

Exhibits/Experiments
- Salt crystals, tweezers, conductometer, distilled water (Would we notice the increase in conductivity in tap water or seawater?)
- Tap water, seawater ($\frac{1}{2}$ M saline solution)
- Body fat scale (How does the scale calculate body fat percentage from conductivity?)
- Permanganate solution in an electric field (electrophoresis—What does the rate of migration of ions depend on?)

Lab Experiments
- Conductivity of strong and weak electrolytes
- Migration velocity of ions in the electric field

Workshop 12: Electrodes

Question of the day
- How much energy is in a zinc–carbon battery?

Exhibits/Experiments
- Lemon battery made of pencil sharpener (Which everyday metals have the largest potential difference?); sacrificial anode
- Fuel cell, battery, accumulator (What is different (anode/cathode/positive pole/minus pole) during charging/discharging)?
- Water electrolyzer (How much voltage is needed for water decomposition?)
- Zinc and copper salt solution (difference between spontaneous and reversible?)
- Oxford Electric Bell (How does the electrochemical experiment that has been going on since 1841 work?)

Lab Experiments
- Open circuit voltage of Galvanic cells
- Hittorf's transfer number

13.4   Links and QR Codes to the Multimedia Courses (Figs. 13.21 and 13.22)

Fig. 13.21 Multimedia Laboratory Course PhysicalChemistry^{LAB} (https://www.ili.fh-aachen.de/goto.php?target=grp_790369)

Fig. 13.22 Multimedia course PhysicalChemistry[101] ("Physical Chemistry in a Nutshell" http://www.PhysChemBasics.de)

13.5   List of Abbreviations

$[A]$	thermodynamic concentration of the substance A (depending on the substance: molarity c, partial pressure p, or mole fraction x)
A	Area
A	Free Energy (HELMHOLTZ Energy)
A	Frequency factor
a	VAN-DER-WAALS constant of the internal pressure
B	2nd virial coefficient
b	Molality
b	VAN-DER-WAALS constant of the covolume
c	Molarity
C_p	isobaric heat capacity (enthalpy capacity)
C_V	isochoric heat capacity (energy capacity)
E_A	Activation energy
D	Diffusion coefficient
E	Electric field
e	Elementary charge
F	FARADAY constant
h	PLANCK's constant
I	Ionic strength
i	VAN'T HOFF factor
J	Mass flow density
k	Reaction rate constant
K_a	Acid constant

k_B	BOLTZMANN constant
k_{eb}	ebullioscopic constant
K_{eq}	thermodynamic equilibrium constant
k_H	HENRY's constant
K_K	KOHLRAUSCH's constant
k_{kr}	cryoscopic constant
$\langle \lambda \rangle$	Average free path length
μ_{J-T}	JOULE–THOMSON coefficient
m	Mass
M	Molar mass
n	Amount of substance
N_A	AVOGADRO constant
n_e	electrochemical coefficient
p	Pressure
Q	Charge
R	ideal gas constant
r	Reaction rate
T	Temperature
T_c	Critical temperature
T_{fus}	Solidification temperature
T_i	Inversion temperature
T_{vap}	Boiling temperature
t	Time
u	Mobility
v	Drift speed
V	Volume
$\langle v \rangle$	Average velocity
$\overline{V}$	molar volume
x	Location coordinate
x	Mass fraction (mole fraction) in the condensed phase
$\langle x \rangle$	Medium displacement
y	Location coordinate
y	Mass fraction (mole fraction) in the gas phase
z	Charge number
Z	Real gas factor (compression factor)
$\langle z \rangle$	Average impact frequency
α	Degree of dissociation
α	coefficient of thermal expansion
γ	Surface tension
η	Dynamic viscosity
θ	Edge angle
κ	Compressibility
λ	Thermal conductivity
μ	Chemical potential (partial molar free enthalpy)

ν	Decay number
ρ	Density (mass concentration)
σ	Impact cross section (interaction cross section)
τ	Shear stress
ϕ	Volume share
χ	Flory – Huggins Coefficient
κ	Specific conductivity
Λ	Molar conductivity
F	Force
L	Length
H	Enthalpy
S	Entropy
G	Free enthalpy (GIBBS energy)
U	Inner energy
q	Heat
w	Work
w_{pV}	(Pressure) Volume work
q_{rev}	reversible heat
Π	Osmotic pressure
π_T	Internal pressure
$\Delta_f H^\circ$	Standard enthalpy of formation
$\Delta_r H^\circ$	Standard reaction enthalpy
$\Delta_{fus} H$	Melting enthalpy
$\Delta_{vap} H$	Evaporation enthalpy
$\Delta_r S^\circ$	Standard reaction entropy
$\Delta_r G^\circ$	Standard reaction affinity (impetus)
$\overline{U}$	Molar internal energy
Ω	thermodynamic probability

13.6 Constants and Units

Gas constant	$R = 8.314\,\frac{J}{mol\,K}$
Avogadro – Konstante	$N_A = 6.022 \cdot 10^{23}\,\frac{1}{mol}$
Boltzmann – Konstante	$k_B = 1.381 \cdot 10^{-23}\,\frac{J}{K}$
Faraday – Konstante	$F = 9.6485 \cdot 10^4\,\frac{C}{mol}$
Elementary charge	$e = 1.6022 \cdot 10^{-19} C$

Units

Energy $J = kPa \cdot L = Pa \cdot m^3 = V \cdot A \cdot s$
Print $1\ bar = 100\ kPa$
        $1\ atm = 760\ Torr = 101.3\ kPa$

13.7   Bond Enthalpies (Table 13.2)

Table 13.2 Bond enthalpies in kJ/mol (single bond/double bond/triple bond arranged one below the other)

	H	C	N	O	S	F	Cl	Br	I
H	−436								
C	−413	−346							
		−602							
		−835							
N	−386	−305	−167						
		−615	−418						
		−887	−945						
O	−463	−358	−201	−146					
		−799	−607	−498					
		−1072							
S	−347	−272			−226				
F	−565	−485	−283	−190	−284	−155			
Cl	−432	−339	−192	−218	−255	−253	−242		
Br	−366	−285		−201	−217	−249	−216	−193	
I	−299	−213		−201		−278	−208	−175	−151

13.8 Thermodynamic Data (Table 13.3)

Table 13.3 Molar mass, enthalpy of formation, normal entropy, and free enthalpy of formation of some inorganic and organic substances

	M in g/mol	$\Delta_f H°$ in kJ/mol	$S°$ in J/(mol K)	$\Delta_f G° (\mu°)$ in kJ/mol
Ag(s)	107.87	0.0	42.6	0.0
AgCl(s)	143.32	−127.1	96.2	−109.8
Ag$_2$O(s)	231.74	−31.1	121.3	−11.2
Al(s)	26.98	0.0	28.3	0.0
Br$_2$(l)	159.82	0.0	152.2	0.0
Br$_2$(g)	159.82	30.9	245.5	3.1
CaO(s)	56.08	−635.1	39.7	−604.0
Ca(OH)$_2$(s)	74.09	−986	83	
CaCO$_3$(s)	100.09	−1212.0	92.9	−1128.8
Cl$_2$(g)	70.91	0.0	223.1	0.0
Cl$^-$(aq)	35.45	−167.2	56.5	−131.2
C(s)(gra.)	12.01	0.0	5.74	0.0
C(s) (dia.)	12.01	1.9	2.4	2.9
C(g)	12.01	716.68	158.1	671.3
CO(g)	28.01	−110.62	198.12	−137.2
CO$_2$(g)	44.01	−393.77	213.86	−394.4
CO$_2$(aq)	44.01	−413.8	117.6	−386.0
H$_2$CO$_3$ (aq)	62.03	−699.7	187.4	−623.1
CH$_4$(g)	16.04	−74.81	186.26	−50.7
C$_2$H$_2$(g)	26.04	226.7	200.9	209.2
C$_2$H$_4$(g)	28.05	52.3	219.6	68.2
C$_2$H$_6$ (g)	30.07	−84.7	229.6	−32.8
C$_3$H$_6$(g)	42.08	20.4	267.1	62.8
C$_6$H$_6$(l)	78.12	49.0	173.3	124.3
C$_6$H$_6$(g)	78.12	82.9	269.3	129.7
C$_6$H$_{12}$(l)	84.16	−156.0		26.8
C$_6$H$_{14}$(l)	86.18	−198.7	204.3	
C$_6$H$_5$CH$_3$(g)	92.14	50.0	320.7	122.0
C$_7$H$_{16}$(l)	100.21	−224.4	328.6	1.0
C$_8$H$_{18}$(l)	114.23	−249.9	361.1	6.4
i-C$_8$H$_{18}$(l)	114.23	−255.1		
C$_{10}$H$_8$(s)	128.18	78.5		
CH$_3$OH(l)	32.04	−238.7	126.8	−166.3
CH$_3$OH(g)	32.04	−200.7	239.8	−162.0
C$_2$H$_5$OH(l)	46.07	−277.7	160.7	−174.8
C$_2$H$_5$OH(g)	46.07	−235.1	282.7	−168.5
HCOOH(l)	46.03	−424.7	129.0	−361.4
CH$_3$OOH(l)	60.05	−484.5	159.8	−389.9
CH$_3$CHOO-Et(l)	88.11	−479.0	259.4	−332.7

(continued)

Table 13.3   (continued)

	M in g/mol	$\Delta_f H°$ in kJ/mol	$S°$ in J/(mol K)	$\Delta_f G°$ $(\mu°)$ in kJ/mol
HCHO(g)	30.03	−108.6	218.8	−102.5
$CH_3CHO(l)$	44.05	−192.3	160.2	−128.1
$CH_3CHO(g)$	44.05	−166.2	250.3	−128.9
$\alpha\text{-D-}C_6H_{12}O_6(s)$	180.16	−1274.0		
$\beta\text{-D-}C_6H_{12}O_6(s)$	180.16	−1268.0	212.0	−910.0
$C_{12}H_{22}O_{11}(s)$	342.3	−2222.0	360.2	−1543.0
CuO(s)	79.54	−157.3	42.6	−129.7
$CuSO_4(s)$	159.6	−771.4	109.0	−661.8
$CuSO_4 \cdot 5H_2O(s)$	249.68	−2279.7	300.4	−1879.7
Fe(s)	55.85	0.0	27.3	0.0
$I_2(s)$	253.81	0.0	116.1	0.0
$I_2(g)$	253.81	62.4	260.7	19.3
$H_2(g)$	2.016	0.0	130.684	0.0
H(g)	1.01	217.94	114.7	203.3
$H^+(aq)$	1.01	0.0	0.0	0.0
$H_2O(l)$	18.02	−285.84	69.9	−237.1
$H_2O(g)$	18.02	−241.83	188.72	−228.6
$H_2O_2(l)$	34.02	−187.8	109.6	−120.4
HCl(g)	36.46	−92.3	186.9	−95.3
HCl(aq)	36.46	−167.2	56.5	−131.2
$H_2S(g)$	34.08	−20.6	205.8	−33.6
Mg(s)	24.31	0.0	32.7	0.0
MgO(s)	40.31	−601.7	26.9	−569.4
$N_2(g)$	28.01	0.0	191.5	0.0
N(g)	14.01	470.6	153.3	455.6
NO(g)	30.01	90.3	210.8	86.6
$N_2O(g)$	44.01	82.1	219.9	104.2
$NO_2(g)$	46.01	33.2	240.1	51.3
$N_2O_4(g)$	92.01	9.2	304.3	97.9
$N_2O_5(s)$	108.01	−43.1	178.2	113.9
$N_2O_5(g)$	108.01	11.3	355.7	+ 115. 1
$NH_3(g)$	17.03	−46.1	192.5	−16.5
$NH_3(aq)$	17.03	−80.3	113.3	−26.5
$NH_4NO_3(s)$	80.04	−365.6	151	
Na(s)	22.99	0.0	51.2	0.0
Na(g)	22.99	107.3	153.7	76.8
$Na^+(aq)$	22.99	−240.1	59.0	−261.9
NaOH(s)	40	−425.6	64.5	−379.5
NaCl(s)	58.44	−411.2	72.1	−384.1
$Na_2CO_3(s)$	105.99	−1130.7	135	
$NaHCO_3(s)$	84.01	−947.7	102	
$O_2(g)$	31.999	0.0	205.0	0.0
O(g)	15.999	249.17	161.06	231.7

(continued)

Table 13.3   (continued)

	M in g/mol	$\Delta_f H°$ in kJ/mol	$S°$ in J/(mol K)	$\Delta_f G°$ ($\mu°$) in kJ/mol
O_3(g)	47.998	142.7	238.9	163.2
OH^-(aq)	17.007	−230.0	−10.8	−157.2
S(s. α)	32.06	0.0	31.8	0.0
S(g)	32.06	278.81	167.82	238.25
SO_2(g)	64.06	−296.8	248.2	−300.2
SO_3(g)	80.06	−395.7	256.8	−371.1
Zn(s)	65.37	0.0	41.6	0.0
Zn(g)	65.37	130.7	161.0	95.1
Zn^{2+}(aq)	65.37	−153.9	− 112. 1	−147.1
Pb(g)	207.19	195.0	175.4	161.9

13.9   Gas Properties (Table 13.4)

Table 13.4   Critical temperatures and pressures, 2nd virial coefficients, van-der-Waals parameters, and Henry's constants of some gases

	p_c in MPa	T_c in K	$\overline{V}_c$ in mL/mol	B in mL/mol bei 0 °C	a in Pa m^6 mol^{-2}	b in mL/mol	K_H in GPa bei 20 °C
Ar	4.8	151	75	−22	0.14	32	4.02
CO_2	7.4	304	94	−142	0.396	42.69	0.165
CH_4	4.6	191	99	−54	0.23	43	3.97
He	0.2	5	58	12	0.003	23	14.5
H_2O	21.8	647	56		0.55	31	
N_2	3.4	126	90	−11	0.14	39	8.68
Ne	2.7	44	42	10			54.7
NH_3	11.1	406	73		0.42	46	0.235
O_2	5.0	155	78	−22	0.14	32	4.4

13.10 ANTOINE Equation and Parameters (Table 13.5)

Table 13.5 Antoine-
Parameter of some fluids

	A	B in °C	C in °C
CH_4	5.7367	389.927	265.99
H_2O	7.19621	1730.63	233.426
N_2	5.6194	255.6778	265.55
O_2	5.8163	319.011	266.7
C_2H_5OH	7.2371	1592.864	226.184
CH_3OH	7.20587	1582.271	239.726
C_2H_6	5.9276	656.401	255. 99
CH_3COCH_3	6.24204	1210.595	229.664
C_3H_8	5.9546	813.199	247.99
i-C_3H_7OH	8.00319	2010.33	252.636
C_6H_6	6.00477	1196.76	219.161
$C_6H_5CH_3$	6.07577	1342.31	219.187
C_6H_{12}	5.97636	1206.47	223.136

$$\log\left(\frac{p_A^*}{kPa}\right) = A - \frac{B}{C + T(\text{in}\,°C)}$$

13.11 Ionic Conductivities (Table 13.6)

Table 13.6 Molar limiting conductivities and ionic mobilities

Ion	H^+	Li^+	Na^+	K^+	SO_4^{2-}	OAc^-	OH^-	Cl^-
$\lambda_\infty\,/\frac{mS\,m^2}{mol}$	34.96	3.87	5.01	7.35	16.0	4.09	19.9	7.63
$u_\infty\,/10^{-8}\frac{m^2}{V\,s}$	36.23	4.01	5.19	7.62	16.6	4.34	20.64	7.91

13.12 Electrochemical Series (Table 13.7)

Table 13.7 Electrochemical series and standard redox potential

Oxid. form/red. form	Electron transfer reaction	E_{redox}^{0} in V
MnO_4^-, H^+/Mn^{2+}, H_2O	$MnO_4^-(aq) + 8\ H^+(aq) + 5\ e^- \rightleftharpoons Mn^{2+}(aq) + 4\ H_2O(l)$	+1.51 @ pH = 0
Cl_2/Cl^-	$Cl_2(g) + 2\ e^- \rightleftharpoons 2\ Cl^-(aq)$	+1.36
O_2, H^+/H_2O	$O_2(g) + 4\ H^+(aq) + 4\ e^- \rightleftharpoons 2\ H_2O(l)$	+1.23 @ pH = 0 +0.82 @ pH = 7
TEMPO-Radikal		≈ 0.95
Ag^+/Ag	$Ag^+(aq) + e^- \rightleftharpoons Ag(s)$	+0.80
O_2, H_2O/OH^-	$O_2(g) + 2\ H_2O(l) + 4\ e^- \rightleftharpoons 4\ OH^-(aq)$	+0.40 @ pH = 14 +0.82 @ pH = 7
Galvinoxyl-Radikal		≈ 0.28
Cu^{2+}/Cu	$Cu^{2+}(aq) + 2\ e^- \rightleftharpoons Cu(s)$	+0.34
$AgCl/Ag$, Cl^-	$AgCl(s) + e^- \rightleftharpoons Ag(s) + Cl^-(aq)$	+0.22
$H^+/H2$	$2\ H^+(aq) + 2\ e^- \rightleftharpoons H_2(g)$	0.00 @ pH = 0 −0.42 @ pH = 7
Fe^{3+}/Fe	$Fe^{3+}(aq) + 3\ e^- \rightleftharpoons Fe(s)$	−0.04
Pb^{2+}/Pb	$Pb^{2+}(aq) + 2\ e^- \rightleftharpoons Pb(s)$	−0.13
Sn^{2+}/Sn	$Sn2+(aq) + 2\ e^- \rightleftharpoons Sn(s)$	−0.14
Fe^{2+}/Fe	$Fe2+(aq) + 2\ e^- \rightleftharpoons Fe(s)$	−0.44
Zn^{2+}/Zn	$Zn2+(aq) + 2\ e^- \rightleftharpoons Zn(s)$	−0.76
H_2O/H_2, OH^-	$2\ H_2O(l) + 2\ e^- \rightleftharpoons H_2(g) + 2\ OH^-(aq)$	−0.83 @ pH = 14 −0.42 @ pH = 7
Al^{3+}/Al	$Al^{3+}(aq) + 3\ e^- \rightleftharpoons Al(s)$	−1.66
Mg^{2+}/Mg	$Mg^{2+}(aq) + 2\ e^- \rightleftharpoons Mg(s)$	−2.36
Li^+/Li	$Li^+(aq) + e^- \rightleftharpoons Li(s)$	−3.05

Bibliography

Atkins, P.W. / de Paula, J. (2013) Physical Chemistry, New York: Wiley-VCH.
Brown, T. E. et al (2017) Chemistry: The Central Science, London: Pearson.
Ender, V. (2015) Praktikum Physikalische Chemie, Berlin: Springer.
Engel, T. / Reid, P. (2013) Physical Chemistry, London: Pearson.
Lauth, J. (2016a) Fundamentals of Thermodynamics and Behavior of Gases, Berlin: Springer.
Lauth, J. (2016b) Chemical Thermodynamics, Berlin: Springer
Lauth, J. (2016c) Phase equilibria, Berlin: Springer.
Lauth, J. (2016d) Reaction kinetics, Berlin: Springer.
Lauth, J. (2016e) Electrochemistry, Berlin: Springer
Lauth, J. / Kowalczyk, J. (2015) Thermodynamik, Berlin: Springer
Lauth, J. / Kowalczyk, J. (2016) Einführung in die Physik und Chemie der Grenzflächen und Kolloide, Berlin: Springer.